3D Digital Design in Ergonomics and Human Factors

3D Digital Design in Ergonomics and Human Factors is the definitive guide to understanding how 3D software impacts the practice of ergonomics and human factors and how it can be utilized successfully in a variety of different settings. It covers interdisciplinary areas, including ergonomics and human factors, 3D digital design, sustainable digital human anatomical design through Open-Source Software (OSS), and advanced technologies in design. It helps readers at any skill level in 2D and 3D design to increase their competency in this ever-growing field of study.

Written in an inclusive, jargon-free way, the book covers the significance of 3D digital design for ergonomics and human factors. It includes an explanation of the structural features of 3D polygonal-mesh modeling and 3D solid modeling (Computer Aided Design—CAD). Within digital OSS, the modeling of anatomical digital humans, integration of AI tools, and advancements in ergonomics, MoCap, and bioengineering for inclusive healthcare are presented in detail. Technologically effective digital OSSs are featured with which the modeling of anatomical digital human, the development of ergonomics and motion capture (MoCap), and ergonomics and bioengineering for inclusive healthcare are possible. Direct useful links to OSS 2D and 3D software and add-ons for expanding the capabilities of digital modeling are presented, and file formats and their extensions receive significant coverage.

This modern and timely book will appeal to students, academics, scientists, and professionals associated with 3D digital design, ergonomics and human factors, digital human modeling, bioengineering, healthcare, information technology, workplace safety, education, and proponents of OSS for 2D and 3D design. It provides readers with the necessary digital tools for their activities and needs by giving real, successful examples from practice.

Human Factors in Design, Engineering, and Computing

Series Editors: Tareq Ahram and Waldemar Karwowski

This series focuses on research and development efforts intended to promote the comprehensive integration of people and technological systems. To help foster integration of human factors research with emerging technologies, this series will focus on novel methodologies, design tools and solutions that advance our understanding of the nature of human factors and ergonomics with intelligent technologies and services. It will address all aspects of human factors and human-centered design, with a particular emphasis on applications of emerging technologies, computing, Artificial Intelligence and Systems. The series offers a multidisciplinary platform for researchers and practitioners alike, discussing emerging issues in the field of human factors engineering systems, with a special focus on (but not limited to) computing and AI-based technologies. It will be useful for researchers, senior graduate students, graduate students, and professionals in different domains including human factors, engineering design, systems science and engineering, and ergonomics.

Human Factors in Engineering
Manufacturing Systems, Automation, and Interactions
Edited by Beata Mrugalska, Tareq Ahram and Waldemar Karwowski

Human Factors in Computing
Social Dimension and Artificial Intelligence-Based Technologies
Edited by Tareq Ahram and Waldemar Karwowski

Human Factors and Emerging Technologies
Integrating People and Technological Systems
Edited by Tareq Ahram and Waldemar Karwowski

Human Factors in Intelligence, Technology and Analytics
Edited by Tareq Ahram and Waldemar Karwowski

Human-Technology
Blending Artificial Intelligence, Computing, and Intelligent Design
Edited by Tareq Ahram and Waldemar Karwowski

Converging Minds
The Creative Potential of Collaborative AI
Aleksandra Przegalinska and Tamilla Triantoro

3D Digital Design in Ergonomics and Human Factors
Tihomir Dovramadjiev

For more information about this series, please visit: www.routledge.com/Human-Factors-In-Design-Engnineering-and-Computing/book-series/%20CRCHFIDEC

3D Digital Design in Ergonomics and Human Factors

Tihomir Dovramadjiev

CRC Press
Taylor & Francis Group
Boca Raton London New York

CRC Press is an imprint of the
Taylor & Francis Group, an **informa** business

Designed cover image: image created by author on MakeHuman OSS software

First edition published 2025
by CRC Press
2385 NW Executive Center Drive, Suite 320, Boca Raton FL 33431

and by CRC Press
4 Park Square, Milton Park, Abingdon, Oxon, OX14 4RN

CRC Press is an imprint of Taylor & Francis Group, LLC

ISBN: 978-1-032-50315-8 (hbk)
ISBN: 978-1-032-51061-3 (pbk)
ISBN: 978-1-003-40092-9 (ebk)

DOI: 10.1201/9781003400929

Typeset in Times
by SPi Technologies India Pvt Ltd (Straive)

Access the Support Material: www.Routledge.com/9781032503158

Contents

Preface

3D Digital Design in Ergonomics and Human Factors is the definitive guide to understanding how 3D software impacts the practice of ergonomics and human factors and how it can be utilized successfully in a variety of different settings. It covers interdisciplinary areas including ergonomics and human factors, 3D digital design, sustainable digital human anatomical design though Open-Source Software (OSS), and advanced technologies in design and others, and will assist any reader to increase their personal qualities and competence in the field. This book explores the intersection of advanced technologies, including artificial intelligence (AI) and real-time networks, with 3D digital design, demonstrating how these tools can be harnessed to solve complex ergonomic challenges in real time. By integrating AI-driven tools and real-time resources, designers can now access dynamic, adaptive solutions that enhance precision, efficiency, and creativity in ergonomic design. The title covers the significance of 3D digital design for ergonomics and human factors and includes an explanation of the structural features of 3D polygonal-mesh modeling and 3D solid modeling (Computer Aided Design—CAD). Technologically effective digital OSSs are featured with which the modeling of anatomical digital human and the development of ergonomics and motion capture (MoCap) and ergonomics and bioengineering for inclusive healthcare are possible. The title includes direct links to OSS 2D and 3D software and add-ons for expanding the capabilities of digital modeling and significantly covers file formats and their extensions. It provides readers with the necessary digital tools for their activities and needs by giving real, successful examples from practice.

The real implementation of 3D design, digital tools, and ergonomic and human factors solutions is brought to life in this book through examples and resources from successfully completed international projects and real-world research applications. These practical implementations illustrate how cutting-edge technologies and methodologies have been applied to solve real-world challenges, providing readers with actionable insights and proven strategies.

This modern and timely book will appeal to students, academics, scientists, and professionals associated with 3D digital design, ergonomics and human factors, digital human modeling, bioengineering, healthcare, information technology, transport, education, and proponents of OSS for 2D and 3D design.

About the Author

Tihomir Dovramadjiev is a professor associated with the Department of Industrial Design at the Technical University of Varna (TUV), Bulgaria. He is the President of the Bulgarian Association of Ergonomics and Human Factors (BAEHF), affiliated with the International Ergonomics Association (IEA) and the Federation of European Ergonomics Societies (FEES). He holds a master's degree in industrial engineering and received his PhD in Ergonomics and Industrial Design in 2012 from TUV. Within the framework of the TUV, he is a leading lecturer of the disciplines: "Advanced Technologies in Design"; "Basics of Computer Design"; "Multimedia Design"; "3D Digitalization" (Bachelors); and "Eco Design" (Masters). He has participated in over 100 scientific publications for prestigious academic publishers. Additionally, he participates in a number of editorial boards and scientific conference committees such as AHFE, IHSI, IHSED, CCCE, ARIN, MLT, and others. He also participated in notable successful international projects as "HEDY Life in the AI Era" and "ErgoDesign—Improving digital skills for Ergonomics and Bioengineering Innovations for inclusive Health Care."

1 Essence of Ergonomics and Human Factors

Ergonomics, also known as Human Factors and Ergonomics (HFE eq. E/HF), is the scientific discipline concerned with understanding the interactions among humans and other elements of a system, applying theories, principles, data, and methods to optimize human well-being and overall system performance.

The International Ergonomics Association (IEA) defines ergonomics as "the science of work," derived from the Greek words ergon (work) and nomos (laws). HFE encompasses three main domains: Physical Ergonomics, which focuses on human anatomical, anthropometric, physiological, and biomechanical characteristics; Cognitive Ergonomics, which deals with mental processes such as perception, memory, reasoning, and motor response; and Organizational Ergonomics, which aims to optimize sociotechnical systems, including their organizational structures, policies, and processes. The practice of HFE is multidisciplinary, integrating knowledge from various fields to address systemic issues using a holistic, user-centric approach. This involves considering physical, cognitive, sociotechnical, organizational, and environmental factors and their interactions. HFE principles are rooted in socio-technical values, emphasizing humans as assets, technology as a tool to assist humans, promotion of quality of life, respect for individual differences, and responsibility to all stakeholders. Effective HFE requires a broad understanding of different areas and a participatory approach, involving consultation with specialists in various domains. The HFE perspective includes both microergonomics (design of procedures, context, and tools) and macroergonomics (work organization, technology, communication, and feedback). HFE contributes to safe, sustainable systems through a systematic, iterative, design-driven process that optimizes performance and well-being. Stakeholders in HFE include system influencers (e.g., governments, regulators), decision-makers (e.g., employers, managers), experts (e.g., HFE specialists, engineers, psychologists), and actors (e.g., employees, product users). HFE is vital for the economic health of organizations, enhancing worker well-being, capability, and sustainability, maximizing performance, and reducing costs associated with productivity losses, quality deficiencies, and employee turnover. HFE design principles lead to better employee performance and improved business outcomes, demonstrating that HFE is not just beneficial but essential for effective work system design and organizational success [1–7].

The IEA accepts the classification of correlation aspects to HFE as shown in Figures 1.1 and 1.2.

Through the "Essence of Ergonomics and Human Factors," leading organizations like the IEA and its networks [8], such as Federation of European Ergonomics Societies (FEES [9]), Asian Council on Ergonomics and Design (ACED [10]), Federation of Brazilian, Russian, Indian, Chinese, and South African HF/E Societies

DOI: 10.1201/9781003400929-1

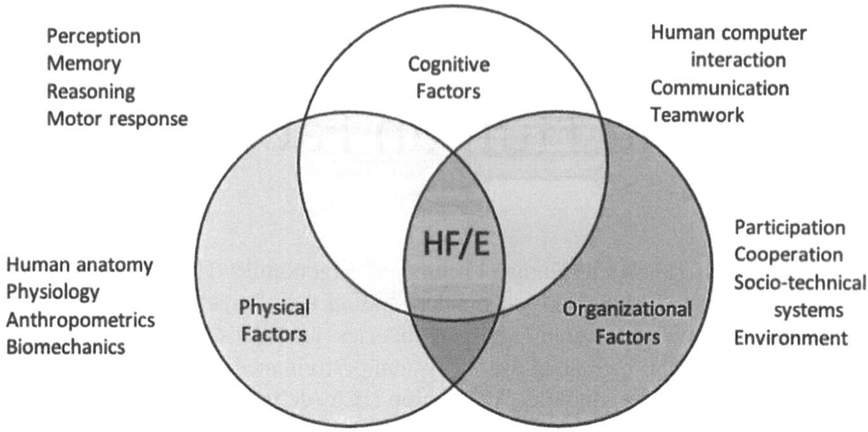

FIGURE 1.1 International Ergonomics Association (IEA): Domains of E/HF https://iea.cc/about/what-is-ergonomics/ [7].

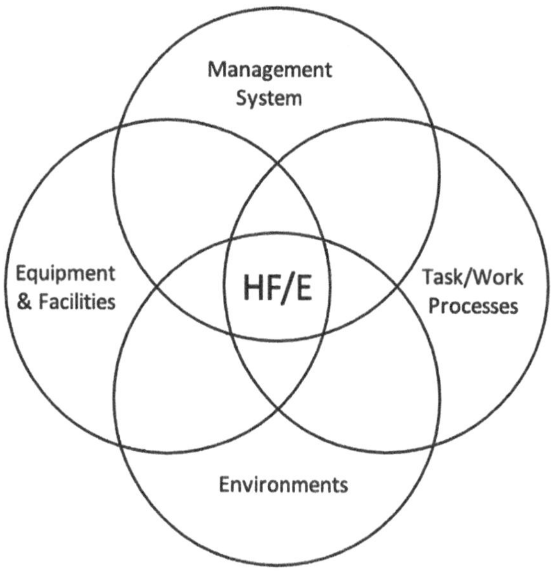

FIGURE 1.2 International Ergonomics Association (IEA): Value of HFE in the world of work https://iea.cc/about/what-is-ergonomics/ [7].

(BRICSplus [11]), La Unión Latinoamericana de Ergonomía (ULAERGO [12]), The South East Asian Network of Ergonomics Societies (SEANES, [13]), ErgoAfrica [14], and related federated societies [15], cover the current social and technological advancements, ensuring continuous improvement and progress. They align their goals with modern needs and future developments, fostering a symbiotic relationship between technology and human expertise in digital era. By doing so, they not only address present challenges but also pave the way for future innovations, enhancing the quality of life and societal progress on a global scale. The logo of the IEA is shown in Figure 1.3 [16].

The IEA plays a crucial role in advancing the field of HFE globally. The IEA aims to elaborate and advance HFE science and practice, expanding its application to improve quality of life. The IEA's mission includes developing effective communication and collaboration between federated societies, advancing HFE at an international level, and enhancing its contribution to society. The IEA's strategic policies focus on engaging stakeholders, promoting HFE science, technology, and practice, reinforcing education and professional standards, strengthening relationships with external partners, and maintaining a future-focused approach. The IEA supports its mission through a robust organizational structure, including the Council, Executive Committee, and various technical committees, ensuring that the field of ergonomics continues to evolve and meet the needs of a dynamic global world. By integrating these efforts with advanced digital design and AI, the field of ergonomics is well-positioned to lead innovations that enhance human well-being and productivity in Industry 5.0 and beyond.

The integration of 3D digital design in HFE has taken place in the transition from Industry 4.0 to Industry 5.0. During Industry 4.0, there was widespread adoption of Computer-Aided Design (CAD), Computer-Aided Manufacturing (CAM), and Computer-Aided Engineering (CAE) systems. These systems facilitated collaboration among designers and engineers, streamlining the design process and improving product quality. CAD allowed for the creation of detailed digital models, CAM converted these models into manufacturing instructions, and CAE enabled virtual testing and analysis of product performance. The Computer-Integrated Manufacturing (CIM) systems integrated CAD/CAM/CAE technologies with production processes, enhancing manufacturing efficiency. In Industry 5.0, the role of Human Factors and Artificial Intelligence (AI) is particularly noteworthy, with both serving as two sides of the same coin. Human Factors ensure that technology is aligned with human needs, preferences, and capabilities, while AI enhances decision-making processes, automates tasks, and augments human capabilities. Together, Human Factors and AI can be viewed as a collaborative hybrid intelligence, embodying the synergistic integration of human expertise and machine intelligence. This concept, known as Hybrid Intelligence Interaction (HII), plays an important role in Industry 5.0 and Society 5.0.

The HII framework emphasizes the collaboration and interaction between humans and intelligent systems, fostering a symbiotic relationship where each complements

FIGURE 1.3 International Ergonomics Association (IEA): Official logo https://iea.cc/ [16].

the strengths of the other. Human Factors guide the development of AI systems, ensuring ethical and human-centric design principles, while AI enhances human decision-making and productivity. This interaction leads to innovative solutions that prioritize human well-being, sustainability, and societal progress. By leveraging CAD/CAM/CAE, CIM systems, 3D digital design, Human Factors, and AI within the HII framework, organizations can further refine ergonomic solutions, enhance human–machine interactions, and prioritize human well-being. This integration creates a collaborative environment where technology empowers individuals to innovate and succeed in an evolving industrial environment.

REFERENCES

[1] International Ergonomics Association (IEA): https://iea.cc/about/what-is-ergonomics/
[2] Bridger, R. S. (2018). *Introduction to Human Factors and Ergonomics*, 4th Edition. Boca Raton, FL: CRC Press.
[3] Read, G. J. M., Salmon, P. M., Goode, N., & Lenné, M. G. (2018). A sociotechnical design toolkit for bridging the gap between systems-based analyses and system design. *Human Factors and Ergonomics in Manufacturing & Service Industries*, 28(6), 327–341.
[4] Principles and Guidelines for HF/E Design and Management of Work Systems. (2019). Joint Document by IEA and the International Labour Organization (ILO).
[5] Wilson, J. R. (2014). Fundamentals of systems of ergonomics/human factors. *Applied Ergonomics* (45), 5–13.
[6] Dul, J., Bruder, R., Buckle, P., Carayon, P., Falzon, P., Marras, W. S., Wilson, J. R., & van der Doelen, B. (2012). A strategy for human factors/ergonomics: Developing the discipline and profession. *Ergonomics*, 55(4), 377–395. DOI: 10.1080/00140139.2012.661087
[7] Hendrick, H. W. (2003). Determining the cost-benefits of ergonomics projects and factors that lead to their success. *Applied Ergonomics*, 34, 419–427.
[8] International Ergonomics Association (IEA): Networks https://iea.cc/leadership/networks/
[9] Federation of European Ergonomics Societies (FEES) http://www.ergonomics-fees.eu/
[10] Asian Council on Ergonomics and Design (ACED) https://aced.asia/
[11] International Ergonomics Association (IEA): Federation of Brazilian, Russian, Indian, Chinese, and South African HF/E Societies https://iea.cc/member/federation-of-brazilian-russian-indian-chinese-and-south-african-human-factors-ergonomics-societies-bricsplus/
[12] La Unión Latinoamericana de Ergonomía (ULAERGO http://www.ulaergo.com/
[13] International Ergonomics Association (IEA): The South East Asian Network of Ergonomics Societies (SEANES) https://iea.cc/member/the-south-east-asian-network-of-ergonomics-societies-seanes/
[14] International Ergonomics Association (IEA): ErgoAfrica https://iea.cc/member/ergoafrica/
[15] International Ergonomics Association (IEA): Federated societies https://iea.cc/leadership/federated-societies/
[16] International Ergonomics Association (IEA): Official logo https://iea.cc/

2 Significance of 3D Digital Design for Ergonomics and Human Factors

One of the best open and accessible platforms for working, researching, and getting science-based data in real time online is OPEN DESIGN LAB [1]. To illustrate design and ergonomic research, the platform offers a rich set of digitally developed people (anthropometric data are scientifically validated). The OPEN DESIGN LAB provides a diverse range of anthropometric data sets to support ergonomic and human factors research. This includes the ANSUR II, a comprehensive survey conducted in 2012 that extends to include US Military reservists, offering updated and broad measurements compared to the ANSUR I survey from 1988, which focused solely on active-duty personnel. Additionally, the platform offers data from the NHANES 2015–2018 and NHANES 2011–2014, segments of the US National Health and Nutrition Examination Survey specifically related to adult anthropometry, providing critical insights into the body measurements of the civilian population. For synthesized data needs, the platform includes Proportionality Constants for estimating body segment lengths based on stature, offering a straightforward method for generating body size and shape data. An Unweighted NHANES 2011–2014 data set is available, allowing for analysis without the need for statistical weights, facilitating more straightforward data manipulation and application [2]. The OPEN DESIGN LAB's Multivariate Design Tools offer a comprehensive suite of resources for ergonomic and human factors design, enabling the visualization and application of anthropometric data in multifaceted ways. The Multivariate Accommodation Calculator 2.0 allows users to visualize various anthropometric dimensions simultaneously via a parallel coordinates plot, facilitating the estimation of accommodation across multiple measures. The Manikin Fetcher enables the selection of Digital Human Models that best fit specified anthropometric parameters, with options to download these models in various CAD file formats for further design use. For practical applications, the Virtual Fit Test spreadsheet offers a downloadable tool for conducting Virtual Fit Trials tailored to the most commonly used measures for designing seated and standing workstations for US civilian populations. The original Multivariate Accommodation Calculator employs sliders to set limits on multiple anthropometric dimensions, calculating accommodation both individually and collectively. The Proportionality Constants tool simplifies the estimation of body segment lengths based on stature. Lastly, Data Explorer II provides an updated platform for exploring and downloading anthropometric data from various populations, including US civilians, ANSUR, ANSUR II, and NHANES, supporting in-depth research and design work [3].

DOI: 10.1201/9781003400929-2

Figure 2.1 showcases a study utilizing the OPEN DESIGN LAB Multivariate Accommodation Calculator 2.0, which allows for the exploration and download of anthropometric data from several key sources [4].

Figures 2.2 (M) and 2.3 (F) present a study of OPEN DESIGN LAB Proportionality Constants, a tool that estimates body segment lengths based on overall stature, providing a simplified approach to synthesizing data on body size and shape. The background information explains that these proportionality constants represent the average ratios of specific body segment lengths to overall height, derived from the work of Drillis and Contini (1966 https://www.yumpu.com/en/document/read/37732232/drills-contini-1966-body-segment-parameters#google_vignette). Although these ratios are widely used, their application can be problematic because individual segment ratios can vary significantly within a population. For instance, while "shoulder height" has a strong correlation with stature (R2 = 0.93), "shoulder breadth" does not (R2 = 0.15). This tool, therefore, is useful for exploring relationships between body segments and stature, but is not recommended for precise design work. The tool reports R2 and RMSE values for each segment, derived from regression analyses on ANSUR data, offering insights into the accuracy of these estimations. For further details on the limitations and applications of proportionality constants, the study references Fromuth and Parkinson [5].

Table 2.1 presents data from the OPEN DESIGN LAB, focusing on the selection of Digital Human Models (DHMs) or "manikins" based on specific anthropometric measurements. Users can filter through a collection of these models to find ones that closely match their desired criteria, such as height, weight, or other body dimensions. Once a suitable model is identified, it can be downloaded in various CAD file formats, enabling integration into different design or engineering applications. The example highlighted includes 3D digital models of male and female, showcasing the lab's capability to provide detailed and accurate representations of human body types for diverse design needs [6].

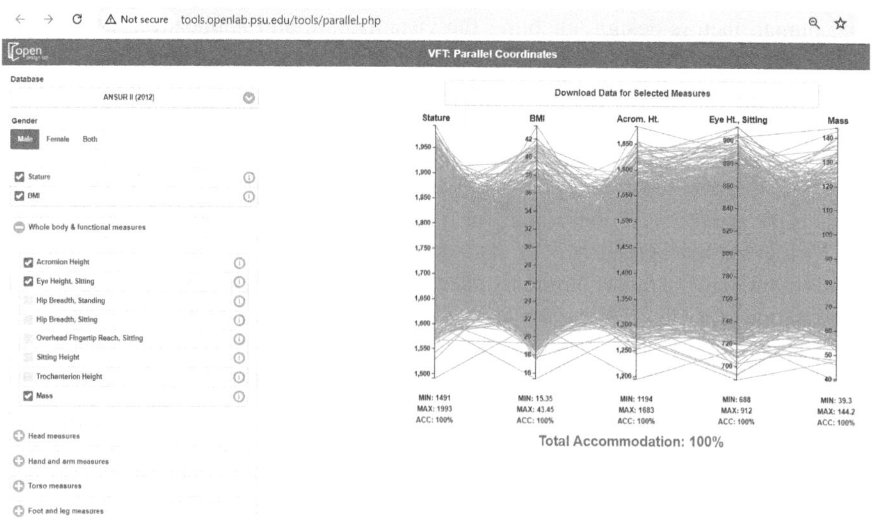

FIGURE 2.1 OPEN DESIGN LAB Multivariate Accommodation Calculating Study (Parameters: Male; Stature; BMI; Acromion Height; Eye Height Sitting; Mass), https://tools.openlab.psu.edu/tools/parallel.php [4].

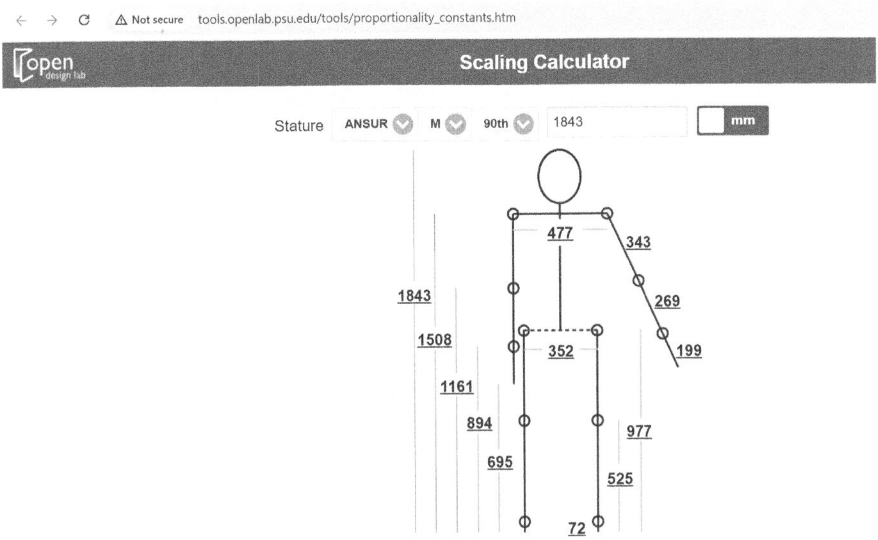

FIGURE 2.2 OPEN DESIGN LAB Proportionality Constants Study of Male ANSUR 90 Percentile, https://tools.openlab.psu.edu/tools/proportionality_constants.htm [5].

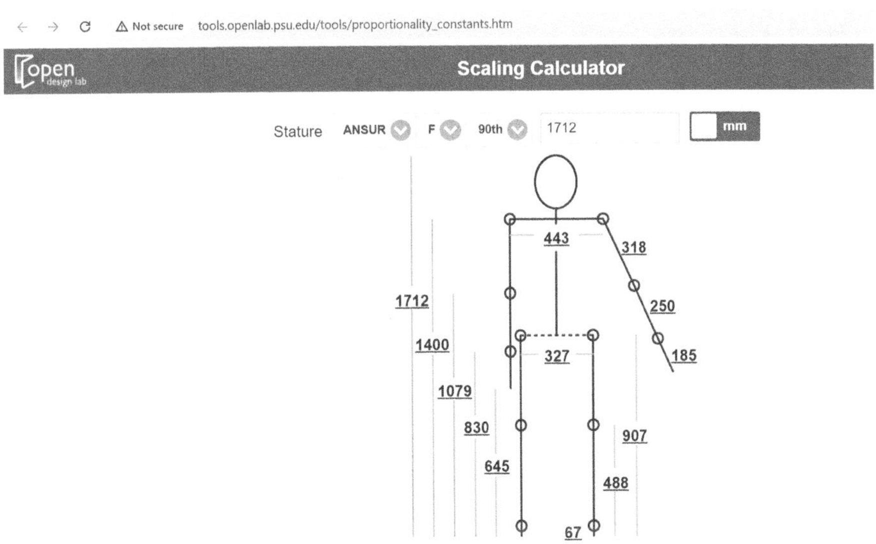

FIGURE 2.3 OPEN DESIGN LAB Proportionality Constants Study of Female ANSUR 90 Percentile, https://tools.openlab.psu.edu/tools/proportionality_constants.htm [5].

This integration of digital design technology has transcended conventional boundaries, fostering a deeper understanding of ergonomic principles and enhancing the effectiveness of human-centric design across various domains, including education, healthcare, industry, and transportation. In education, 3D digital design facilitates immersive learning experiences, empowering students to engage with ergonomic

TABLE 2.1
OPEN DESIGN LAB: Digital Human Models (DHMs), "manikins" Based on Specific Anthropometric Measurements [6]

Posture:	OPEN DESIGN LAB Choose specific anthropometric measurements to locate Digital Human Models (or "manikins") from collection that closely match the criteria. Download the selected manikins in different CAD file formats [6] https://tools.openlab.psu.edu/tools/fetcher.php	3D Digital Model: Male Blender software
Posture: Standing Neutral Mm: 1854	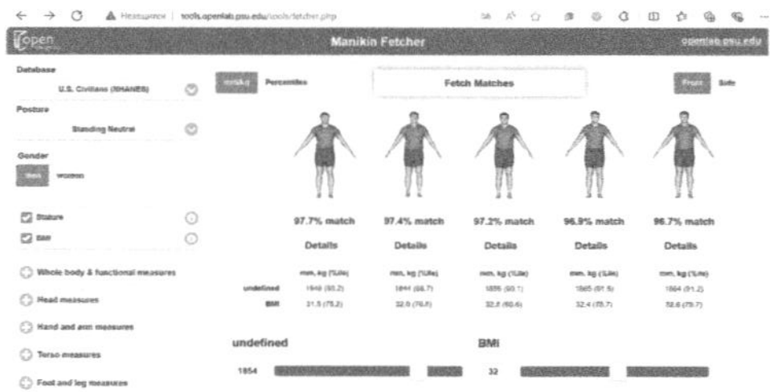	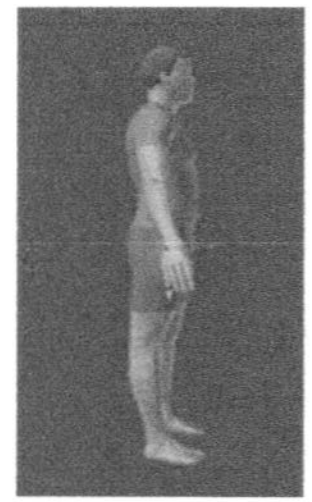
Posture: Standing Working Mm: 1854		
Posture: Seated Typing Mm: 1854		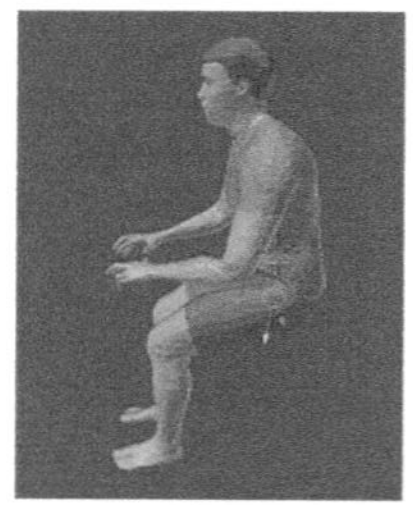

Posture:	OPEN DESIGN LAB Choose specific anthropometric measurements to locate Digital Human Models (or "manikins") from collection that closely match the criteria. Download the selected manikins in different CAD file formats [6]. https://tools.openlab.psu.edu/tools/fetcher.php	3D Digital Model: FemaleBlender software
Posture: Standing NeutralMm: 1703	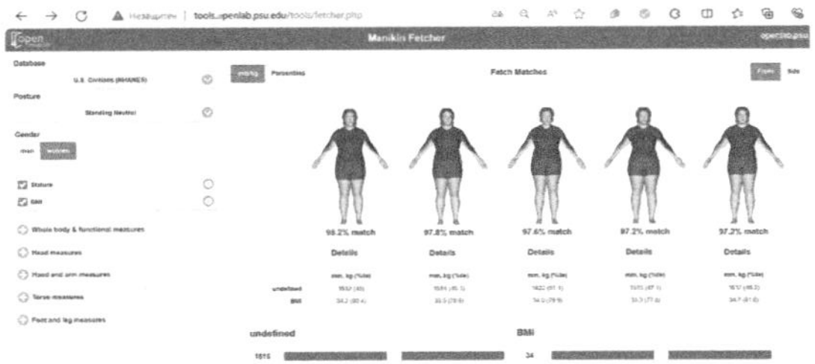	
Posture: Standing Working Pull Mm: 1703		
Posture: Standing Working Push Mm: 1703		

concepts through interactive exploration and practical applications. In healthcare, it drives innovation, allowing healthcare professionals to design customized medical solutions that prioritize patient comfort, safety, and optimal outcomes. Within industrial settings, 3D digital design streamlines manufacturing processes, fosters the development of ergonomic tools and equipment, and cultivates work environments that promote employee well-being, productivity, and satisfaction. In essence, the significance of 3D digital design for ergonomics and human factors lies in its ability to bridge the gap between technology and humanity, fostering innovation, efficiency, and well-being in an ever-evolving world. Importance for the general and advanced 3D digitization in the era of highly enhanced artificial intelligence (AI) is the basic field experience related to conventional design for ergonomics and human factors needs, including: Education, Healthcare, and Safety Workplaces.

2.1 IN EDUCATION

In education, 3D digital design improves the learning experience by providing immersive and interactive tools that enable students to grasp ergonomic concepts with greater depth and understanding. Through simulations and practical applications, students can engage with complex ergonomic principles, fostering a hands-on learning approach that enhances retention and comprehension.

A notable international project in this context is the European project "HEDY Life in the AI Era" (https://lifeintheaiera.eu/) [7]. HEDY aims to provide massive open online courses (MOOC) for free and open access, focusing on Life in Artificial Intelligence. This project is crucial as AI represents the 4th industrial revolution, and HEDY's goal is to be a free and accessible source of information about this digital technology, which is becoming an integral part of our lives. The HEDY project offers several valuable resources in 7 languages [8]:

HEDY BOOKLET: A study defining the Hedy position on Life in the AI Era, organizing AI features, positive impacts, and associated risks. It identifies challenges, opportunities, and expected impacts on business, governance, work, education, culture, and health.

HEDY TOOLKIT: A collection of influential audio-visual tools demonstrating how AI could tackle some of the world's most challenging societal issues and the risks to mitigate for AI to realize its full potential.

MOOC: Massive open online courses that are free and accessible to participants, with interactive forums.

HEDY GUIDELINES: Concise, easy-to-read guidelines for the results described above, plus a glossary with key terms on AI-related topics. It explains the nature of the Hedy MOOC learning resources and shows exciting ways to use the Toolkit assets in public events.

The project is developed by an international consortium from: BULGARIA: Bulgarian Association of Ergonomics and Human Factors (BAEHF) [9]; GERMANY: ACEEU GmbH [10]; HUNGARY: Coordinator: OBUDAI EGYETEM (University) [11]; PORTUGAL: AIDLEARN, Consultoria em Recursos Humanos, Lda. [12]; SPAIN: UNIVERSITAT POLITECNICA DE CATALUNYA [13].

The International Ergonomics Association (IEA) has officially announced the successful completion of the European Union Erasmus + project HEDY Life in the AI Era [14]. The notable scientific results of the HEDY project have been published in a special issue [15].

2.2 IN HEALTHCARE

Within the realm of healthcare, the impact of 3D digital design is transformative, leading to the development of customized medical solutions that prioritize patient comfort, safety, and overall well-being. From personalized prosthetics to ergonomic medical devices, digital design technology enables healthcare professionals to tailor treatments to individual needs, resulting in improved outcomes and quality of care.

An important project with high international scope is the ErgoDesign (https://ergodesigner.eu/, [16, 17]), officially titled "ErgoDesign - Improving Digital Skills for Ergonomics and Bioengineering Innovations for Inclusive Health Care," which is a pioneering project in the healthcare sector that develops a comprehensive, multidisciplinary, and transdisciplinary e-learning course. This course trains higher education students to become adept designers and producers of healthcare implants, with a particular focus on meeting the needs of individuals with special requirements. The overarching goal of ErgoDesign is to integrate ergonomic and bioengineering innovations into the fields of implantology and prosthetics by utilizing new digital applications and providing specialized training in their use. The project has several specific objectives, including increasing students' knowledge in ergonomics and bioengineering, enhancing their digital skills for designing and producing 3D printed medical products such as dental and orthopedic implants, prostheses, and exoskeletons. Additionally, it deepens the understanding necessary for creating healthcare implants tailored for individuals with physical and mental special needs. The project also focuses on expanding trainers' expertise in software and digital tools, boosting their capacity to adapt these tools for educational purposes, thereby enhancing the employability of students in the medical implants sector. ErgoDesign establishes a vibrant community of practice centered on ergonomics and bioengineering. Throughout the project's implementation, several significant results are developed. Firstly, a specialized software toolkit designed for the healthcare sector is created, presenting various digital tools to enhance the digital competencies of higher education lecturers, academics, and students. Secondly, the project establishes the ErgoDesign Curriculum and Course Design Principles, defining the multidisciplinary curriculum's foundational topics, learning outcomes, and duration. Thirdly, a collaborative platform is launched to provide a space for discussion, training, and the virtual community of practice, fostering stakeholder engagement. Fourthly, ErgoDesign Training Materials are developed to address knowledge and skill gaps among students, combining theoretical education with practical exercises. Finally, a comprehensive handbook and e-learning course are produced to facilitate the replication of the ErgoDesign course by other organizations, ensuring the project's sustainability and broader impact. The consortium is responsible for the ErgoDesign project: Poznan University of Technology (PUT)—Poznan, Poland—General Coordinator; Technical University of Varna (TUV)—Varna, Bulgaria; Óbuda University (ÓU)—Budapest, Hungary; National Technical University of

Athens (NTUA)—Athens, Greece; Technická univerzita v Košiciach (TUKE)—Kosice, Slovakia; ValueDo—Florence, Italy. Substantive partners: Universities AP: University of West Attica School of Engineering (GR); University of Presov (SL); University of Zielona Góra (PL); Lodz University of Technology (PL); Organization Center of Registration of European Ergonomics, University of Bordeaux (FR). It is the only body endorsed to certify professional ergonomists in Europe. Networks and ICSTED: BAEHF, Polish Ergonomics Society, an interdisciplinary organization for the development and popularization of Ergonomics in Poland, Inclusion and Innovation Technology Center, representing engineers' and engineering knowledge within the largest Hungarian disability organization, ICSTED [18]. Some of the notable scientific research results of the ErgoDesign project had been published in IETI [19, 20].

2.3 IN SAFETY WORKPLACES

The "Strengthening Workplace Safety through Applied Ergonomics in the Digital Age" (SAFE-D) project (https://www.dsgi.hu/safe-d/ [21, 22]), running from April 1, 2024, to March 31, 2026, is an ERASMUS+ initiative aimed at enhancing workplace safety through the integration of ergonomic principles in the context of modern digital workplaces. The DSGI Ergonómiai Mérnökiroda Kft. (Hungary) in collaboration with the Bulgarian Association of Ergonomics and Human Factors and the Cappadocia Innovation Institute (Turkey), this project targets enterprises. It seeks to improve employee health, efficiency, and satisfaction by addressing specific needs in workplace safety. The project's objectives include conducting a needs assessment to tailor an ergonomics curriculum, offering workshops and mentorship programs for practical application, and effectively managing and disseminating project outcomes to broaden the reach of safe working conditions. The SAFE-D project underscores the importance of creating healthy and safe work environments and aims to involve transnational partners to maximize its impact, reflecting a shared commitment to promoting workplace safety across borders. The project is supported by the European Union, although the views expressed within it are solely those of the participants and do not necessarily represent the positions of the EU or the European Education and Culture Executive Agency (EACEA).

In real time, in order to advance news related to ergonomics and human factors in correlation with digital innovations, 3D design, AI, and other related fields, leading international scientific conferences accumulating current developments and achievements of scientists from all over the world are organized annually. These are, for example, congresses and conferences such as: IEA Congresses [23, 24], Applied Human Factors and Ergonomics (AHFE) [31], the International Conference on Human Systems Engineering and Design (IHSED) [25], the Intelligent Human Systems Integration (IHSI) [26], the International Conference on Human Interaction and Emerging Technologies (IHIET) [27], International Conference on Computer and Communication Engineering (CCCE) [28], International Conference on NLP & AI (NLPAI) [29], Materials Science, Engineering and Technology [30], ErgoWork [31], International Seminar of Ergonomics [32], and other.

In the field of science books, journals, and special issues, the scientific advantages show companies with editions like: Taylor and Francis [33], Ergonomics & Human

Factors [34], Springer [35], Ergonomics and Human Factors for a Sustainable Future, Current Research and Future Possibilities [36], Elsevier [37], Advances in Human Factors/Ergonomics Book series [38]. The International Engineering and Technology Institute (IETI). IETI is a nonprofit organization dedicated to promoting innovations in science, engineering, and technology globally, with no membership fees. Beyond traditional disciplines, IETI encompasses financial engineering, financial econometrics, financial statistics, mathematics, statistics, social sciences, and business. As a multidisciplinary public research organization, IETI aims to advance science, engineering, and technology worldwide. Currently, IETI boasts over 5,000 members, including engineers, doctors, and professors from various fields. More than 100 professors have been recognized as IETI Distinguished Fellows and Fellows, among whom are renowned experts and laureates of prestigious awards like the Nobel Prize, Turing Award, Fields Medal, Wolf Prize, and John Bates Clark Medal. Related IETI journals are: IETI Transactions on Engineering Research and Practice ISSN 2616-1699 [39]. IETI Transactions on Ergonomics and Safety ISSN 2520-5439 [40]. The list with various important journals are: EnPress Journals [41]; Science Publishing Group [42]; MDPI Mathematics and Computer Programming in 2D and 3D Open Source Software [43]; MDPI Mathematics, Computer Programming, and Artificial Intelligence in 2D and 3D Software [44]; both in Mathematics, ISSN 2227-7390 [45], Journal Rank: JCR - Q1 (Mathematics) / CiteScore - Q1 (MDPI General Mathematics); Occupational Ergonomics, Human Factors and Safety: Theory, Application, and Practice [46]; Human Factors and Ergonomics: Bridging the Gap between Research and Practice in Occupational Safety and Health [47]. Advances in Ergonomics, Human Factors and Safety for Occupational Health [48]. Occupational Health: Safety Management, Safety Human Factors and Ergonomics [49]; Human-Computer Interaction, Ergonomics and Health [50], all in MDPI International Journal of Environmental Research and Public Health [51], Journal Rank: CiteScore - Q1, State-of-the-Art in Human Factors and Interaction Design [52] and Ergonomics and Human Factors in Transportation Systems [53], both in MDPI Applied Sciences [54], Journal Rank: JCR - Q2 (Engineering, Multidisciplinary) / CiteScore - Q1 (General Engineering), Ergonomics and Sustainability [55] and Advances in Ergonomics and Safety [56] both in MDPI Safety [57] Journal Rank: CiteScore - Q2 (Safety Research), AI in Human Factors and Ergonomics [58] in MDPI AI Journal Rank: CiteScore – Q2 (Artificial Intelligence) [59]; Human Factors in Complex Systems [60] in MDPI Systems [61] Journal Rank: JCR - Q2 (Social Sciences, Interdisciplinary) / CiteScore - Q2 (Modeling and Simulation) and other.

To elevate and develop the significance of 3D digital design for ergonomics and human factors on both international and European levels, International Summer Universities and specialized workshops are organized. These events provide platforms for professionals, researchers, and students to collaborate, share knowledge, and explore advancements in 3D digital design. They facilitate the exchange of ideas and best practices, fostering innovation and enhancing the application of 3D digital design in ergonomics and human factors. Through these educational and collaborative initiatives, participants gain insights into cutting-edge technologies and methodologies, driving progress in the field and promoting global standards. With success had been completed: The 34th edition of the Summer University on Ergonomics

[62, 63]. Co-organized by the Austrian Ergonomics Society, the Hungarian Ergonomics Society, ERGOWORK from Romania and the BAEHF. This edition was unique in its thematic approach, exploring ergonomics beyond the prevention of musculoskeletal disorders in line with the EU-OSHA "Lighten the Load" campaign. The program included technical presentations, best practice showcases, and specific language tracks, such as Bulgarian focusing on AI and ergonomics, Hungarian on office and sedentary work, Polish on mental workload, and Romanian on service ergonomics. Universities collaborated through the CEEPUS Ergonomics and Human Factors Regional Educational Network, facilitating international dialogue and advancing ergonomic practice through interdisciplinary exchange. Detailed information and registration were available on their event page and registration page. The 4th PhD/master student online workshop series, organized by the Ergonomics and Human Factors Regional Educational CEEPUS Network, was held at Óbuda University, Budapest, on November 21, 2023. The professional objective of the Educational CEEPUS Network was to explore the legal requirements of a joint safety-related doctoral program and seek solutions for shared mentoring possibilities. This workshop was part of the first phase of implementing the cooperative program. The series aimed to open opportunities for young researchers to initiate cross-border research, share ideas, and publish scientific papers. The event featured both in-person and online participation, fostering a collaborative environment for emerging scholars [64]. The virtual workshop titled "Technology for Co-working on Health and Active Life," endorsed by FEES and supported by AAL Europe AAL (European program funding innovation that keeps us connected, healthy, active, and happy into our old age, [65]) took place on June 13th, 2024. This comprehensive two-hour session addressed the evolving health care systems in European regions, emphasizing the integration of innovative processes, competence improvement strategies, and technology applications. The workshop aimed to foster interdisciplinary collaboration among healthcare, ergonomics, occupational health, and technology research communities, highlighting the importance of a human-centered approach in AI development (Title: Virtual workshop: Technology for co-working on health and active life [66]). Key topics covered included the introduction to the new format of interdisciplinary collaboration, focusing on current challenges in health care and presenting a novel approach: community health care, co-working on health, and promoting active life. The session delved into discussing agile development concepts and the critical need for ergonomics needs, ISO standards, Eco Design aspects, data structures, particularly open data, to support integrated processes and other. Participants explored the concept of integrated learning and joint cognitive systems (JCS), emphasizing their relevance to ergonomics and the management of change in healthcare organizations. Organized in collaboration with the main Austrian company KOHS, BAEHF, and the Ergonomics and Human Factors Regional Educational CEEPUS Network [67], the workshop addressed actual demands in healthcare for secondary users, highlighting the applications of technology to meet these needs. It also focused on the health of primary users and health care workers, examining the challenges faced by health care organizations and occupational health professionals. By fostering a collaborative environment, the workshop sought to enable co-working on health and enhance the quality of life through the integration of technology and innovative processes.

The significance of 3D digital design for ergonomics and human factors is deeply intertwined with our modern lives. In an era where technology permeates every aspect of daily existence, the integration of 3D digital design into ergonomics and human factors represents a keypoint advancement. This integration fosters vital relationships between technological innovation, academic pursuits, and scientific activities. Such collaborations, especially within joint ergonomics and human factors networks and organizations, are of utmost importance as people navigate a future increasingly influenced by AI and sophisticated technologies. 3D digital design acts as a mirror of our 3D reality, reflecting the complexities and nuances of our physical world in a digital format. It enables the creation of detailed and accurate digital representations of real-world environments, tasks, and interactions. By doing so, 3D digital design provides a powerful tool for ergonomists and human factors professionals to analyze, evaluate, and enhance human interactions with systems and environments.

Incorporating 3D digital design into ergonomics and human factors not only allows for the replication of regular and specific physical activities but also facilitates the simulation of these activities in virtual environments. This capability is crucial for creating duplicate clones of real-life scenarios, enabling the testing and refinement of designs in a risk-free and controlled manner. By simulating physical activities and interactions, researchers and designers can identify potential issues, optimize designs, and ensure that systems and environments are tailored to meet human needs effectively. The implementation of 3D digital design is essential in the context of AI. As AI continues to integrate into everyday life, the need for ergonomically sound and human-centered design becomes increasingly critical. 3D digital design enables the development of AI systems that are more intuitive, user-friendly, and aligned with human capabilities and limitations. By leveraging 3D digital tools, designers can create AI interfaces and interactions that enhance user experience and improve overall system performance. The collaborative efforts of technology developers, academics, and scientists are advancing 3D digital design within the field of ergonomics and human factors. These partnerships facilitate the exchange of knowledge, the development of best practices, and the establishment of standards that guide the effective use of 3D digital tools. Joint networks and organizations play a crucial role in fostering these collaborations, promoting interdisciplinary research, and driving innovation in the field.

The significance of 3D digital design for ergonomics and human factors is a base component of modern life, enabling the creation of precise digital replicas of physical activities and environments. By integrating 3D digital design into the field, people and digital experts and designers can enhance human–system interactions, improve safety and performance, and ensure that new technologies, including AI, are designed with a deep understanding of human needs. As people continue to embrace the digital age, the importance of these collaborative efforts will only grow, paving the way for a future where technology and human factors are seamlessly integrated.

REFERENCES

[1] OPEN DESIGN LAB: Edu https://www.openlab.psu.edu/
[2] OPEN DESIGN LAB: Data https://www.openlab.psu.edu/data/
[3] OPEN DESIGN LAB: Tools https://www.openlab.psu.edu/tools/

[4] OPEN DESIGN LAB: Tools / Parallel https://tools.openlab.psu.edu/tools/parallel.php
[5] OPEN DESIGN LAB: Proportionality Constants Study https://tools.openlab.psu.edu/tools/proportionality_constants.htm
[6] OPEN DESIGN LAB: Manikin Fetcher https://tools.openlab.psu.edu/tools/fetcher.php
[7] "HEDY Life in the AI Era" (HEDY PROJECT): Official webpage https://lifeintheaiera.eu/
[8] HEDY PROJECT: Results https://lifeintheaiera.eu/results/
[9] Bulgarian Association of Ergonomics and Human Factors (BAEHF) https://baehfofficial.wixsite.com/baehf
[10] GERMANY: ACEEU GmbH https://www.aceeu.org/
[11] OBUDAI EGYETEM (University) https://uni-obuda.hu/
[12] AIDLEARN, Consultoria em Recursos Humanos, Lda. https://www.aidlearn.pt
[13] UNIVERSITAT POLITECNICA DE CATALUNYA https://www.upc.edu/ca
[14] InternationalErgonomicsAssociation(IEA):EuropeanUnionErasmus+projectHEDYLifein theAIErahttps://iea.cc/completed-european-union-erasmus-project-hedy-life-in-the-ai-era/
[15] The special issue of the Safety and Security Sciences Review "HEDY – Life in the age of AI" was published with the support of the ERASMUS+ partnership cooperation in the higher education sectors project (KA220-HED 0C8D3623). https://biztonsagtudomanyi.szemle.uni-obuda.hu/index.php/home/issue/view/21
[16] ErgoDesign project: official webpage https://ergodesigner.eu/
[17] ErgoDesign project: logo https://ergodesigner.tu-varna.bg/)
[18] ErgoDesign project: International partnership: https://www.facebook.com/profile.php?id=100078144126313&sk=about_privacy_and_legal_info
[19] Tihomir Dovramadjiev (2024). Exploring The Dynamics of Implantology: Insights Into Anatomy and Practical Examples. *IETI Transactions on Ergonomics and Safety (TES)*, Volume 8, Issue 1, 37–48, USA, ISSN 2520-5439, DOI: 10.6722/TES.202404_8(1).0005. https://www.ieti.net/TES/2024V8I1/5.pdf
[20] DIana Pavlova and Tihomir Dovramadjiev (2024). Visualizing Health: The Significance of Medical Imaging and Human Factors. *IETI Transactions on Ergonomics and Safety (TES)*, Volume 8, Issue 1, 1–11, DOI: 10.6722/TES.202404_8(1).0001. https://www.ieti.net/TES/2024V8I1/1.pdf
[21] European project The "Strengthening Workplace Safety through Applied Ergonomics in the Digital Age" (SAFE-D): official webpage https://www.dsgi.hu/safe-d/
[22] European project The "Strengthening Workplace Safety through Applied Ergonomics in the Digital Age" (SAFE-D): BAEHF https://baehfofficial.wixsite.com/baehf/erasmus-project-safe-d
[23] IEA Congresses: IEA 2024 congress https://www.iea2024.com/
[24] Applied Human Factors and Ergonomics (AHFE) https://ahfe.org/
[25] International Conference on Human Systems Engineering and Design (IHSED) https://ihsed.org/
[26] Intelligent Human Systems Integration (IHSI) https://ihsint.org/
[27] International Conference on Human Interaction and Emerging Technologies (IHIET) https://ihiet.org/)
[28] International Conference on Computer and Communication Engineering (CCCE) https://www.ccce.net/
[29] International Conference on NLP & AI (NLPAI) https://nlpai2024.org/index)
[30] Materials Science, Engineering and Technology https://materialsconference.net/
[31] ErgoWork https://www.mpt.upt.ro/cercetare/conferinte/ergowork.html
[32] International Seminar of Ergonomics https://iset.poznan.pl/organizatorzy_en.htm
[33] Taylor and Francis https://www.tandfonline.com/

[34] Ergonomics & Human Factors (https://www.taylorfrancis.com/collections/ergonomics-human-factors/rho019?context=ubx)
[35] Springer https://www.springer.com Ergonomics and Human Factors for a Sustainable Future
[36] Current Research and Future Possibilities https://www.springerprofessional.de/ergonomics-and-human-factors-for-a-sustainable-future/15592038
[37] Elsevier https://www.elsevier.com/
[38] Advances in Human Factors/Ergonomics Book series https://www.sciencedirect.com/bookseries/advances-in-human-factors-ergonomics
[39] IETI Transactions on Engineering Research and Practice ISSN 2616-1699 https://ieti.net/terp/index.html
[40] IETI Transactions on Ergonomics and Safety ISSN 2520-5439 https://ieti.net/tes/index.html
[41] EnPress Journals https://systems.enpress-publisher.com/
[42] Science Publishing Group https://www.sciencepg.com/
[43] MDPI Mathematics and Computer Programming in 2D and 3D Open Source Software https://www.mdpi.com/journal/mathematics/special_issues/Math_Comput_Program
[44] MDPI Mathematics, Computer Programming, and Artificial Intelligence in 2D and 3D Software https://www.mdpi.com/journal/mathematics/special_issues/Math_Comput_Program_AI
[45] MDPI Mathematics, ISSN 2227-7390, https://www.mdpi.com/journal/mathematics
[46] MDPI: Occupational Ergonomics, Human Factors and Safety: Theory, Application, and Practice https://www.mdpi.com/journal/ijerph/special_issues/occupation_safe
[47] MDPI: Human Factors and Ergonomics: Bridging the Gap between Research and Practice in Occupational Safety and Health https://www.mdpi.com/journal/ijerph/special_issues/hfe_occupational_safety_health
[48] MDPI: Advances in Ergonomics, Human Factors and Safety for Occupational Health, https://www.mdpi.com/journal/ijerph/special_issues/WK9LO3PJ5C
[49] MDPI: Occupational Health: Safety Management, Safety Human Factors and Ergonomics https://www.mdpi.com/journal/ijerph/special_issues/Occup_Health_Safety
[50] MDPI: Human-Computer Interaction, Ergonomics and Health https://www.mdpi.com/journal/ijerph/special_issues/ergonomics
[51] MDPI International Journal of Environmental Research and Public Health, https://www.mdpi.com/journal/ijerph
[52] MDPI: State-of-the-Art in Human Factors and Interaction Design https://www.mdpi.com/journal/applsci/special_issues/nteraction_Design
[53] MDPI: Ergonomics and Human Factors in Transportation Systems https://www.mdpi.com/journal/applsci/special_issues/HF4KB3G561
[54] MDPI Applied Sciences https://www.mdpi.com/journal/applsci
[55] Ergonomics and Sustainability https://www.mdpi.com/journal/safety/special_issues/Ergonomics_Sustainability
[56] Advances in Ergonomics and Safety https://www.mdpi.com/journal/safety/special_issues/8P2J7B48W6
[57] MDPI Safety https://www.mdpi.com/journal/safety, Journal Rank: CiteScore - Q2 (Safety Research)
[58] MDPI: AI in Human Factors and Ergonomics https://www.mdpi.com/journal/ai/special_issues/AI_Human_Factors
[59] MDPI AI Journal Rank: CiteScore - Q2 (Artificial Intelligence) https://www.mdpi.com/journal/ai

[60] MDPI Human Factors in Complex Systems https://www.mdpi.com/journal/systems/special_issues/HFCS

[61] MDPI Systems https://www.mdpi.com/journal/systems

[62] 34th edition of the Summer University on Ergonomics: FEES https://ergonomics-fees.eu/node/283

[63] 34th edition of the Summer University on Ergonomics: BAEHF https://baehfofficial.wixsite.com/baehf/about-3

[64] BAEHF: Workshops https://baehfofficial.wixsite.com/baehf/workshops

[65] AAL Europe AAL https://www.aal-europe.eu/

[66] Virtual Workshop: Technology for co-working on health and active life, https://kohs.at/cms/en/2024/04/26/virtual-workshop-technology-for-co-working-on-health-and-active-life/

[67] Ergonomics and Human Factors Regional Educational CEEPUS Network https://sites.google.com/view/ceepusergohf/home

3 Correlation between Physical and Digital Space through Advanced Technologies in Design

The correlation between physical and digital spaces through advanced technologies in design, particularly in the realm of 3D digital design in ergonomics and human factors, is a vital and evolving field. Advanced additive technologies such as 3D scanning, photogrammetry, 3D printing, and Computer Numerical Control (CNC) machining allow the creation of highly accurate and detailed digital replicas of physical environments and objects, which can then be analyzed and manipulated to optimize ergonomic design. For instance, 3D scanning and photogrammetry enable the precise capture of physical dimensions and features, creating a digital twin that serves as a foundation for further analysis and modification. This digital model can be used to identify potential ergonomic issues, simulate various human interactions, and test different design solutions in a virtual space before implementing them in the real world. The 3D printing and CNC machining facilitate the rapid prototyping and production of physical models based on these digital designs. This seamless transition from digital to physical allows designers and engineers to quickly iterate and refine their concepts, ensuring that the final products meet ergonomic standards and user needs. The ability to produce physical prototypes also provides valuable opportunities for user testing and feedback, further enhancing the design process. In the context of ergonomics and human factors, these advanced technologies offer significant benefits. They enable the creation of customized solutions tailored to individual users' needs, enhancing comfort, safety, and performance. For example, 3D scanning can capture the unique contours of a person's body, allowing for the design of personalized ergonomic tools or workspaces. Similarly, photogrammetry can be used to create accurate models of complex environments, aiding in the design of safer and more efficient workflows.

As artificial intelligence and digital innovations become increasingly integrated into our daily lives, the importance of these technologies in bridging the physical and digital worlds will only grow. They not only enhance our ability to understand and improve human interactions with systems and environments but also support the development of more responsive and adaptive design solutions. By leveraging the capabilities of 3D scanning, photogrammetry, 3D printing, and CNC machining, we can create more effective and human-centered designs that improve the quality of life and work. This integration of physical and digital spaces through advanced design technologies represents a significant step forward in the field of ergonomics and

DOI: 10.1201/9781003400929-3

human factors, promising a future where our built environments are better aligned with our physiological and psychological needs.

3.1 3D SCANNERS

3D scanning technology is very important for 3D digital design, particularly within the domains of ergonomics and human factors, by capturing intricate details and dimensions of physical objects and environments to create precise digital models. The scanning process employs various sophisticated techniques such as laser triangulation, structured light, or photogrammetry to gather millions of data points, forming a detailed point cloud that is subsequently processed into a 3D polygonal mesh. This meticulous digital representation is essential for ergonomic analysis, as it allows designers to evaluate how products or environments fit users with high precision. For instance, a 3D scan of a workstation reveals potential ergonomic issues like awkward postures, reach limitations, and insufficient clearance, enabling designers to make data-driven adjustments to enhance comfort, safety, and efficiency. 3D scanning supports the customization of ergonomic solutions, as individual physical characteristics vary widely. By scanning a person's body, designers can create tailor-made products, from office chairs and tools to medical implants, ensuring a perfect fit that boosts comfort and reduces the risk of musculoskeletal disorders. This technology also accelerates prototyping and iterative design processes; digital models can be modified and tested in virtual environments to assess ergonomic impacts before physical prototypes are made, reducing both time and costs associated with traditional methods. 3D scanning bridges the gap between digital and physical spaces, integrating real-world data with virtual simulations to refine ergonomic solutions. This integration is crucial for designing practical and effective workspaces, optimizing layouts, and improving workplace safety by identifying potential hazards. In the medical field, 3D scanners are used to create custom-fitted orthotics and prosthetics, ensuring devices match the exact contours of a patient's anatomy for improved comfort and functionality, and guide surgical planning with accurate anatomical models. The technology also has significant implications for research and education in ergonomics and human factors, providing insights into human biomechanics and movement patterns that inform ergonomic standards and guidelines, while also serving as interactive learning tools in educational settings to illustrate ergonomic principles. 3D scanners provide the technological possibility to capture the precise geometry of objects and human bodies through techniques such as laser triangulation, structured light, and photogrammetry. These scanners function by emitting light or lasers onto a surface and calculating distance based on the time it takes for the light to reflect, a method known as time-of-flight, or by measuring the deformation of a projected light pattern on the surface. They determine positional data by analyzing the angle of reflection, which allows for the accurate mapping of an object's dimensions. The calculation of distance and position involves complex algorithms that process the light or laser data to create highly detailed 3D models. The accuracy of these measurements can be influenced by the surface characteristics of the scanned objects; for instance, black or polished surfaces can pose challenges due to their light-absorbing or reflective properties, respectively. Black surfaces may absorb the emitted light, reducing the amount of reflected signal and making it difficult to detect the geometry accurately. Conversely, polished or reflective surfaces can cause scattering of the light, leading to inaccuracies in the captured data. Despite these challenges, advancements in scanning technology and software algorithms have improved the ability to handle such surfaces. Professional sprays have also been developed to cover black and polished surfaces, thereby aiding the 3D scanning process by creating a matte, uniform surface that reflects light more predictably. This capability is invaluable for ergonomics, as it allows for the comprehensive analysis of human biomechanics and movement patterns, which are essential for designing products and workspaces that enhance comfort, efficiency, and safety. The precise data collected can be used to create personalized ergonomic solutions. 3D scanning technology facilitates the development of ergonomic standards and guidelines by providing robust data on how people interact with various environments and tools. This data can be analyzed to identify trends and common issues, informing the creation of standards that promote better ergonomic practices. The process of 3D digitalization of a human model with the XYZ hand scanner is shown on the official website of the XYZ company [1].

XYZ hand scanner, a series of portable, high-resolution 3D scanners compatible with laptops and desktops, which are utilized to scan humans or objects, generating precise 3D files. This technology is indispensable for capturing real-world objects and preserving them digitally, allowing for extensive analysis and manipulation. The scanners offer multiple scan modes, enabling users to scan large areas, such as a person or a large object. Alternatively, they can switch to modes that provide smaller, high-detail scans, which are ideal for creating detailed, printable models. The accompanying XYZ Scan Handy software facilitates scanning and postediting processes, allowing users to refine their models and export them in various formats, including .stl, .obj, .ply, and .fbx. This versatility is crucial for applications in ergonomics, where detailed 3D models can help design products and environments that fit human needs more effectively, improving comfort, efficiency, and safety [2].

Another high-quality 3D scanner digitalizing human models for ergonomics needs is a Creality CR-Scan Ferret Pro 3D [3]. The Creality CR-Scan Ferret Pro 3D Scanner is an exceptional device tailored for digitizing human models and other objects with outstanding precision, making it indispensable for ergonomics and human factors applications. Its portable and handheld design, coupled with an advanced ASIC chipset and dual-mode scanning, achieves an impressive 0.1mm accuracy. This level of precision is crucial for creating detailed ergonomic models that help in assessing human interactions with products and environments. The scanner is compatible with multiple platforms, including iOS, macOS, Windows 10/11, and Android, offering versatile usability across different devices. It features revolutionary anti-shake technology that stabilizes scans in real-time, effectively countering hand tremors to produce accurate, seamless results, even when used by individuals with unsteady hands. This capability is particularly beneficial for ergonomic assessments and design, where precise human model scanning is essential. The innovative one-shot 3D imaging enhances scan success rates, ensuring highly accurate models, which is vital for detailed ergonomic analysis and human factor studies. The CR-Scan Ferret Pro excels at scanning challenging surfaces such as black or metallic objects without the need for scanning spray, thus simplifying the process and broadening its applicability. Its 24-bit full-color scanning is powered by a high-resolution color camera, capable of capturing intricate details and original colors of scanned objects, including faces, with exceptional realism. The face mapping algorithm further ensures that detailed facial features are accurately captured, producing lifelike 3D models crucial for ergonomic and anthropometric studies. User-friendliness is a key feature, allowing users to start 3D modeling with a single click, which is particularly advantageous for those new to 3D scanning technology. Weighing just 105 grams, its compact design ensures it is highly portable, enabling users to perform 3D scans on the go, whether documenting sculptures, architectural details, or other objects encountered in various settings. The CR-Scan Ferret Pro is optimized for outdoor use, maintaining excellent performance in diverse lighting conditions, including direct sunlight up to 30,000 lux, thanks to its advanced optical technology and intelligent algorithms. This adaptability makes it suitable for outdoor ergonomic assessments and capturing environmental elements relevant to human factors studies.

3.2 PHOTOGRAMMETRY

Photogrammetry, a cutting-edge technology that captures multiple photographs of an object, environment, or person from various angles to create detailed and accurate 3D models, has profoundly transformed the field of 3D digital design, particularly in ergonomics and human factors. By employing the triangulation principle to determine spatial relationships between points in the images, photogrammetry generates precise three-dimensional representations that are invaluable in ergonomics. This technology enables the creation of digital twins of physical environments and human forms, allowing for meticulous analysis and optimization of workspaces to address ergonomic issues such as awkward postures and insufficient clearance, ultimately enhancing comfort, safety, and productivity. Its ability to produce high-fidelity models is also crucial in developing customized ergonomic solutions, where personalized designs can be tailored to the unique dimensions of individual users. For instance, detailed 3D scans of a person's body can be used to craft custom-fitted office furniture, tools, and wearable devices, thus reducing the risk of musculoskeletal disorders and improving overall well-being. The applications extend to various industries, including automotive and aerospace, where photogrammetry informs the ergonomic design of vehicle interiors and aircraft cockpits by capturing the physical dimensions of users to ensure optimal positioning of controls and seating. The photogrammetry facilitates iterative design and virtual testing, enabling rapid prototyping and refinement by allowing designers to simulate different scenarios and evaluate ergonomic interactions in a virtual environment. This integration of physical and digital spaces is essential for developing practical and effective ergonomic solutions, as it bridges the gap between real-world data and virtual simulations. In occupational health, photogrammetry aids in identifying ergonomic risks and designing interventions to improve workplace safety and productivity by scanning entire environments for comprehensive analysis. In the medical field, it supports the creation of custom-fitted medical devices and enhances surgical planning with detailed anatomical models. Educational and research applications are also significant, as photogrammetry helps study human biomechanics and posture, informs ergonomic standards, and creates interactive learning tools. The photogrammetry enriches advanced technologies, like virtual reality (VR) and augmented reality (AR), providing realistic and functional experiences for ergonomic training and real-time guidance for adjustments. This seamless integration of digital and physical realms ensures that ergonomic solutions are both innovative and grounded in real-world application.

3D digital design in ergonomics and human factors integrates advanced photogrammetry software to produce high-quality, accurate 3D models essential for analyzing and improving human-product interaction. The photogrammetry software like AliceVision Meshroom (free software), 3DF Zephyr, Autodesk ReCap, and RealityCapture, are in the leader list and most popular for stakeholders. AliceVision Meshroom offers a comprehensive photogrammetry workflow, utilizing GPUs to balance high quality with performance for scalable projects involving over 1,000 images. Its nodal editor, featuring a caching mechanism, enhances structure-from-motion solutions with new images, while post-processing features include mesh decimation and retexturing capabilities using custom UVs or automatic generation. Meshroom supports high dynamic range (HDR) panorama workflows, aligning multi-bracketing LDR images into HDR, and enables render farm integration, allowing seamless incorporation into larger computational systems to optimize performance based on specific CPU, RAM, and GPU requirements. Autodesk ReCap Pro transforms photographs and laser scans into detailed 3D models, facilitating the creation of point clouds or meshes and promoting collaboration across teams through cloud-based workflows. Users require Autodesk Flex tokens to process ReCap data services, such as Scan to Mesh and ReCap Photo projects, enhancing workflow efficiency and file management in the cloud. 3DF Zephyr excels in automatically reconstructing 3D models from various sensor inputs, offering full control over the 3D reconstruction steps with multi-GPU support for enhanced performance. Its versatility is evident in applications ranging from orthophotos and digital terrain models to computer-aided design (CAD) drawings, sections, paths, and contour lines, along with extensive exporting capabilities. Available in multiple versions, from free to perpetual licenses, 3DF Zephyr caters to different needs and budgets, providing solutions for a wide range of projects. RealityCapture offers a free version for small businesses and educational use, featuring comprehensive capabilities and learning resources. For larger businesses, a paid version includes access to Unreal Engine and Twinmotion, enhancing integration and workflow efficiency. RealityScan, a mobile app under RealityCapture, enables high-fidelity 3D model creation using mobile devices, guided by AR tools that assist in capturing the necessary photos. Features like the cropping tool help remove noise around the object, while the review mode ensures model quality before direct export to Sketchfab for further use in Unreal Engine, Twinmotion, Mesh to MetaHuman, or other post-processing tools. Collectively, these photogrammetry solutions empower ergonomic and human factors professionals to create precise digital models, fostering improved design and analysis of human-product interactions. Table 3.1 shows the leading photogrammetry software and their technical capabilities.

TABLE 3.1
Lead Photogrammetry Software and Their Technical Capabilities [4–7]

Photogrammetry Software	Digitalization Opportunities
AliceVision Meshroom (full free software) [4] https://alicevision.org/	Meshroom allows the creation of custom pipelines, using a fully open-source nodal workflow that can be tailored to specific needs without recreating the entire process, with support available for additional customizations.
3DF Zephyr [5] https://www.3dflow.net/3df-zephyr-photogrammetry-software/	The free version of 3DF Zephyr software includes: Full 3D reconstruction 50 photos limit Single NVIDIA GPU support Limited exporting capabilities Limited editing tools
Autodesk ReCap [6] ReCap Pro: Subscription https://www.autodesk.com/products/recap/overview?term=1-YEAR&tab=subscription&plc=RECAP	Autodesk ReCap Pro helps designers and engineers create high-quality 3D models from photos or laser scans, delivering point clouds or meshes, and enabling efficient cloud-based collaboration and file management.
RealityCapture [7] Free to use For individuals and small businesses that made less than $1 million in revenue in the past 12 months, educational institutions, and students. All RealityCapture features Access to learning materials Community forums https://www.capturingreality.com/realitycapture	RealityCapture is a cutting-edge photogrammetry software known for its speed and efficiency, enabling automatic creation of virtual reality scenes, textured 3D meshes, orthographic projections, georeferenced maps, and more from images and laser scans.

3.3 3D PRINTERS AND CNC MACHINES

3D digital design has profoundly transformed the field of ergonomics and human factors through the advent of 3D printing and CNC machines, both are state-of-the-art technologies in crafting customized, precise, and functionally superior products for individual needs. These advanced manufacturing technologies enable the creation of highly personalized ergonomic solutions that significantly enhance user comfort, safety, and performance. By integrating 3D printing and CNC machining into the design process, designers and ergonomists can rapidly prototype and iterate on ergonomic tools and equipment, addressing unique anthropometric and biomechanical requirements with unprecedented accuracy. For instance, 3D printers excel at producing complex geometries and intricate designs that traditional manufacturing methods cannot achieve, allowing for the creation of custom-fitted office chairs, keyboards, and mouse designs that conform to the precise contours of an individual's body, thereby reducing strain and discomfort during extended use. This bespoke approach is instrumental in preventing work-related musculoskeletal disorders and boosting productivity. Rapid prototyping is a notable advantage of 3D printing, as it allows designers to quickly produce and test multiple iterations of a product, incorporating immediate feedback and adjustments to optimize user comfort and efficiency. For example, in designing an ergonomic tool handle, various shapes and sizes can be rapidly tested to determine the most comfortable and effective design, accelerating the design cycle and reducing costs. CNC machines, renowned for their high-precision capabilities, complement this process by producing components from diverse materials—metals, plastics, and wood—with exacting tolerances, which is crucial for ergonomic design where precision directly impacts usability and safety. CNC machining is particularly valuable in manufacturing ergonomic equipment and fixtures in industrial and medical contexts, such as custom surgical instruments that enhance precision and reduce fatigue or ergonomic components in manufacturing settings that mitigate awkward postures and repetitive strain injuries. The synergy between 3D printing and CNC machining provides a robust framework for ergonomic design, where a product can be prototyped using 3D printing to refine its ergonomic features and then finalized with CNC machining to ensure durability and precision, yielding a high-quality solution that meets both ergonomic and functional demands. This hybrid approach supports the creation of tailored workstations and tools that adapt to individual worker dimensions and preferences, fostering a healthier, more productive work environment. Beyond individual products, these technologies are instrumental in developing ergonomic solutions for complex systems and environments, such as custom-fitted vehicle interiors and cockpit components in the automotive and aerospace industries, which enhance comfort, accessibility, and safety. In the medical field, 3D printing and CNC machining facilitate the creation of custom orthotics, prosthetics, and rehabilitation devices tailored to the unique anatomy of each patient, improving fit, comfort, and functionality, and supporting rehabilitation by enhancing mobility and independence. Educational and research sectors also benefit from these technologies, which enable students and researchers to create and test innovative ergonomic designs, fostering practical learning experiences and advancing knowledge in ergonomics and human factors.

Stratasys, one of the pioneers in 3D printing technology, and SolidWorks CAD software, has propelled 3D digital and physical design to new heights since the first decade of the 21st century by seamlessly integrating SolidWorks' powerful CAD capabilities with Stratasys' advanced Fused Deposition Modeling (FDM) technology, allowing designers to create detailed 3D models in SolidWorks and print them with Stratasys printers, thus streamlining the design-to-prototype process, enhancing precision, and significantly reducing development time and costs [8].

The Stratasys Idea Series makes 3D printing accessible at an individual level, transforming the 3D printer into an "idea engine." This technology enables designers, engineers, and small design teams to bring their concepts to life swiftly and efficiently. It allows for rapid prototyping, functional testing, and iterative refinement, which accelerates the creative process and enhances the quality of the final product. The Stratasys Idea Series, featuring models like the Mojo and uPrint SE™, uPrint SE Plus™, offer an affordable and desktop-friendly solution, effectively integrating into any personal workspace. These printers utilize the proven FDM technology, which builds models layer by layer using ABSplus™ thermoplastic, ensuring strength and durability for rigorous testing and quality feedback. The seamless integration with 3D CAD software, facilitated by tools such as CatalystEX™ and Print Wizard software, streamlines the process from digital design to physical model. The WaveWash® Support Cleaning System further simplifies post-processing by dissolving support materials, leaving behind precise and functional 3D models [9].

In the next years, the collaboration between Stratasys and SolidWorks will deepen with the development of SolidView software from Stratasys Direct Manufacturing. SolidView lets anyone view, measure, markup, and communicate 3D and 2D designs and assemblies using a standard Windows PC. Its intuitive user interface makes it ideal for rapid prototyping, reverse engineering, and general view and markup applications. Extensive cross-sectioning and measurement features make it easy to identify and document design changes. Automatic slide record and playback, along with a complete suite of markup tools, enable full 3D communication of complex designs and assemblies. SolidView adds instant value to existing CAD data by allowing everyone involved in the product development process—engineering, manufacturing, purchasing, Quality Assurance, suppliers, marketing and sales, and even customers—to share the design data on their Windows PC. SolidView reads the following 3D and 2D formats: SLDPRT, SLDASM, STL, VRML, OBJ, DXF, CGM, HPGL, HPGL2, TIFF, GIF, BMP, with options for IGES and VDA import, as well as network licensing. This enhancement will significantly bolster the efficiency and communication in product development, leveraging both Stratasys' and SolidWorks' strengths.

In modern 3D printing, popular materials include polylactic acid (PLA) for its ease of use and biodegradability; thermoplastic polyurethane (TPU) for flexibility and durability; acrylonitrile butadiene styrene (ABS) for strength and heat resistance; polyvinyl alcohol (PVA) as a water-soluble support material; Nylon for its robustness and flexibility; and Breakaway Support for PLA, which offers easy-to-remove support structures. Each material serves distinct purposes, enhancing the versatility and functionality of 3D printing applications. The importance of ergonomics and the designer's choice and experience cannot be overstated, as understanding the properties and best practices for working with each material is crucial for achieving optimal results. A designer's familiarity with these materials significantly impacts the efficiency of the design process, the quality of the final product, and the overall success of the 3D printing project [10].

Modern 3D printers and materials have diversified, enhancing technical capabilities and significantly speeding up the workflow. This evolution in 3D printing technology not only democratizes innovation by keeping ideas under the designer's control but also saves time and resources, fostering a more dynamic and efficient design environment. With the worldwide increase in the availability and capabilities of 3D models and printers by leading companies, there is also a substantial growth in the design and utilization of 3D open printing. This open-source approach to 3D printing empowers a broader community of designers and engineers to collaborate, share, and innovate, driving further advancements in the field and expanding the possibilities for personalized and cutting-edge applications in various industries.

Similar to the rise of 3D printing, CNC machines have also seen significant worldwide growth. Today, there are hybrid machines that combine laser cutting, engraving, 3D printing, and CNC capabilities, exemplified by models like the Snapmaker. Snapmaker, founded in 2016, has quickly gained prominence by enabling users to swiftly turn ideas into reality with its premium multifunction desktop fabricating machines. The company's products, integrating 3D printing, laser engraving, and CNC carving, offer a versatile 3-in-1 functionality that enhances creativity and accessibility. Snapmaker has won multiple international awards and set a record in 2019 as the most crowdfunded 3D printer globally, with 90% of its orders coming from overseas. Committed to becoming a leading technology company, Snapmaker emphasizes research and development to advance digital machining technologies and ensures high-quality production through its dedicated factory. Through its advertising services, Snapmaker promotes its innovative 3D printers, laser engravers, and CNC cutters worldwide, ensuring a superior consumer experience and fostering a global community of creators [11].

EXAMPLE 3.1 THE SIGNIFICANCE OF MEDICAL IMAGING AND HUMAN FACTORS IN DENTISTRY

A good example of high quality 3D design (digital and physical) and role of Human Factors, which are important for healthcare, is a key focus of the ErgoDesign project. The project results cover most state-of-the-art technologies—such as computed tomography (CT), 3D scanning, 3D software, 3D printing, and 3D CNC machining—used in producing dental prostheses. This international consortium focuses on enhancing digital skills for ergonomics and bioengineering innovations. The project underscores the critical role of advanced technologies and anatomical understanding in inclusive healthcare, thereby contributing to the advancement and dissemination of knowledge in the fields of ergonomics and healthcare innovation. CT, which utilizes X-ray technology and sophisticated computer algorithms, produces detailed cross-sectional images that are indispensable in various medical specialties. In dentistry, CT imaging provides exhaustive views of the jaw, teeth, and surrounding tissues, allowing for precise treatment planning, particularly in the placement of dental implants. By visualizing the jawbone in three dimensions, dentists can assess bone density and quality, determine optimal implant locations, and identify oral pathologies like cysts and tumors, thereby enhancing treatment outcomes and patient care. The integration of CT technology in dental practice highlights the significance of ergonomic design and enhanced usability, ensuring that CT scanners are seamlessly incorporated into clinical workflows, thus optimizing efficiency and minimizing errors. This approach not only improves individual patient care but also fosters interdisciplinary collaboration between dental professionals and other healthcare providers [Section 2, sources: 18, 21].

Step-by-step technology and hand design	Process: Creating a dental model involves a multistep process that integrates advanced software and additive technologies, starting with CT scans to capture detailed anatomical data of the jaw. This data is then refined and adjusted during the 3D digital modeling stage in a software environment. Finally, CNC technology is used in the 3D physical modeling phase to produce a precise and detailed physical model. In the next: Handwork. (Impact of Human Factors in conventional design work). Testing construction: through this iterative process, dental professionals evaluate the fit, alignment, and functionality of the model, ensuring it accurately reflects real-world dental scenarios and meets the highest standards of quality and precision in patient care. (Human art experience) Hand painting: This meticulous hand painting process transforms the dental model into a lifelike representation of oral anatomy, enhancing patient communication, treatment planning, and serving educational purposes in dental practice. In the next polishing with paste on the dental model. This meticulous process results in a smooth, lustrous finish, significantly enhancing the model's realism. Completing: Completing the dental model involves the meticulous assembly of its components to achieve the final result	3D (digital & physical) design of dentistry prosthesis ErgoDesign Project experience: http://ergodesigner.eu/ [Section 2, source 18]Source: DIana Pavlova and Tihomir Dovramadjiev. VISUALIZING HEALTH: THE SIGNIFICANCE OF MEDICAL IMAGING AND HUMAN FACTORS. IETI Transactions on Ergonomics and Safety (TES) 2024, Volume 8, Issue 1, 1-11, USA, ISSN 2520-5439, DOI:10.6722/TES.202404_8(1).0001. https://www.ieti.net/TES/2024V8I1/1.pdf [Section 2, source 21]

1: Working with software and additive technologies	Medical imaging	
2: Working with software and additive technologies	3D digital modeling: scanned data of the jaw in software environment(Dental experts pick up the software from various lists of software and approved design protocols)	
3: Working with software and additive technologies	3D physical modeling: CNC(In medical dentistry cases with needs of temporary construction, dental experts use 3D printing and next complete the final 3D physical models with CNC)	
4: Testing the construction of the dentistry example model:	Handwork. Testing the construction of a dental model involves meticulous handwork and assembly to ensure precision and functionality. Starting with the lower jaw bone and implants, each component is carefully positioned and secured to create a stable foundation.	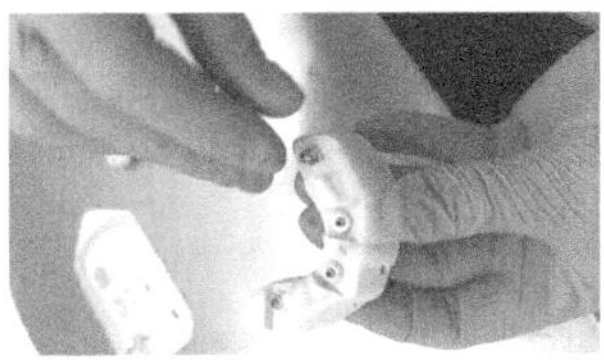
5: Testing construction of dentistry example model: Handwork	Lower jawbone and implants. The process then incorporates teeth, meticulously aligning them with the implants for optimal aesthetics and functionality.	
6: Testing construction of dentistry example model: Handwork	Lower jaw bone, implants, teeth, and fixing components. Finally, fixing components are added to ensure stability and durability.	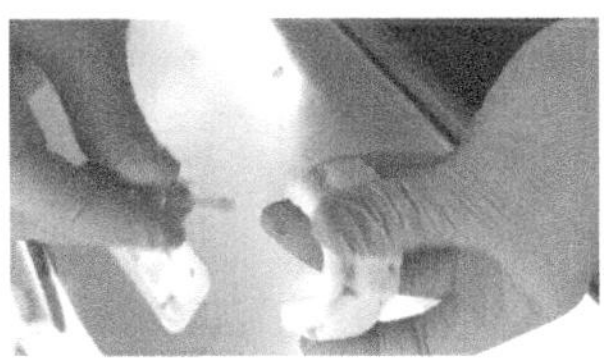

7: Painting of dentistry example model: Handwork	Coloring teeth. Hand painting the dental model requires meticulous attention to detail to achieve lifelike realism and aesthetic appeal. Skilled dental professionals carefully apply paint to the model's surfaces, blending colors and textures to replicate natural tooth enamel.	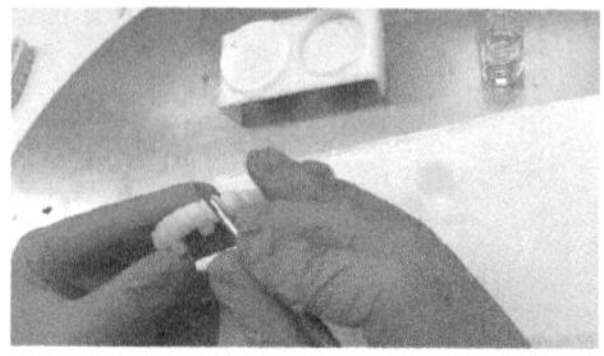
8: Painting of dentistry example model: Handwork	Coloring teeth. They then use intricate shading and highlighting techniques to mimic the nuances of real dentition, ensuring a realistic and visually appealing result.	
9: Polishing with paste, the dentistry example model (PMMA): Handwork	Polishing teeth. Polishing with paste on the dental model involves meticulous refinement to enhance surface texture and achieve a professional finish. Dental experts apply polishing paste to the tooth surfaces using specialized tools and techniques	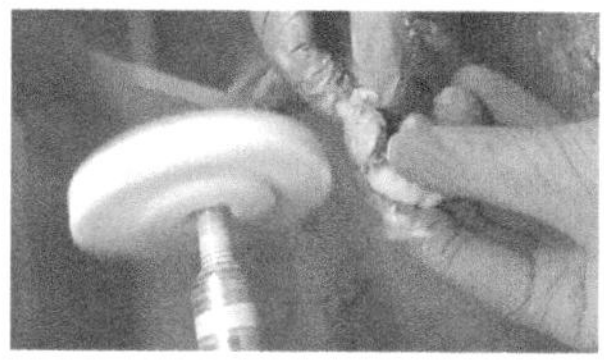
10: Polishing with paste the dentistry example model (PMMA): Hand work	Polishing teeth. Dental experts then meticulously buff and smooth the surfaces to remove imperfections and create a glossy appearance	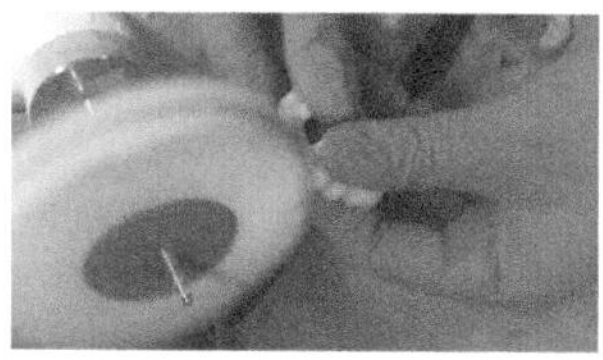
11: Polishing with paste the dentistry example model (PMMA): Handwork	Precision polishing the teeth. Precision polishing includes intricate detailing and careful contouring to ensure uniformity and natural aesthetics across the entire model.	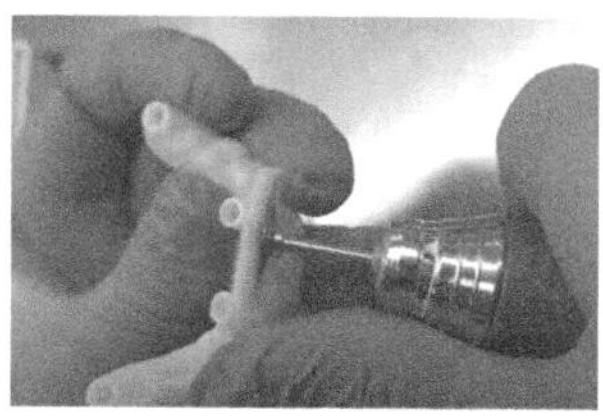
12: Completing the dentistry example model: Assembling	Lower jaw. Completing the dental model involves the meticulous assembly of its components to achieve the final result. The process begins with the careful arrangement and alignment of the lower jaw component.	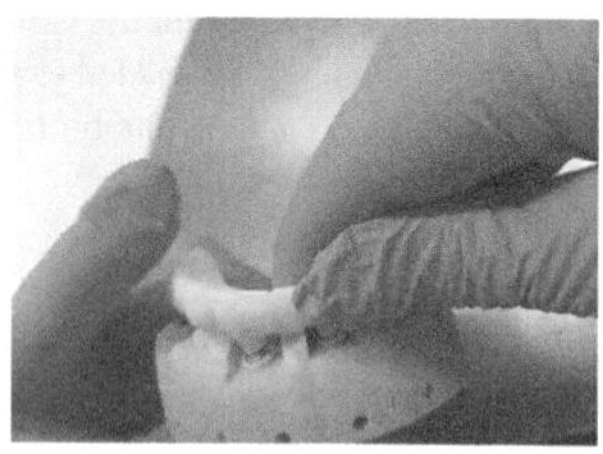

13: Completing the dentistry example model: Assembling	All components. Dental implants, teeth, and fixing components are integrated and secured.	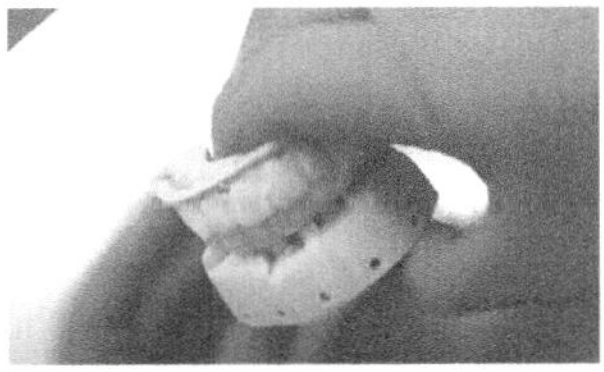
14: Completing the dentistry example model: Assembling	Final ready result. The final stage is the presentation of the fully assembled model.	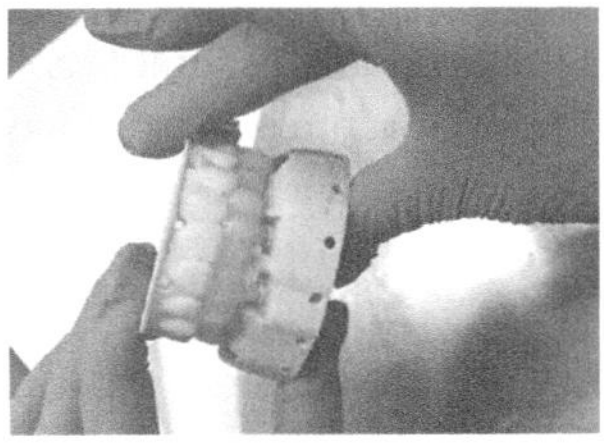

Similarly, in orthopedics, CT imaging has revolutionized the assessment of bone fractures, joint integrity, and musculoskeletal abnormalities. Orthopedic surgeons rely on CT scans to precisely evaluate the extent and location of fractures, aiding in the development of tailored treatment strategies. The detailed images provided by CT scans are crucial for assessing ligament and tendon injuries, cartilage defects, and degenerative joint diseases, thereby facilitating optimal patient outcomes. The ergonomic integration of CT technology in orthopedic practice ensures that imaging devices fit seamlessly into clinical workflows, enhancing both treatment efficacy and patient safety. This interdisciplinary approach underscores a commitment to excellence in orthopedic care. Additive technologies, such as 3D printing and bioprinting, further enhance healthcare innovation by offering personalized solutions tailored to individual patient needs. These technologies enable precise dental reconstructions using digitalization and CNC techniques, as well as advancements in regenerative medicine. The collaboration between clinicians, engineers, and researchers is crucial for navigating the complexities of medical innovation and shaping a future where precision, efficiency, and patient-centric care are prioritized.

EXAMPLE 3.2 PROCESS OF MAKING A LEG PROSTHESIS

The creation of leg prostheses in healthcare exemplifies the importance of integrating medical expertise and the skills of 3D designers with advanced software and additive technologies. This process begins with medical professionals assessing the patient's specific needs and accurately capturing the contours of the healthy and affected areas through precise 3D scanning. The expertise of the 3D designer is crucial at this stage, as they utilize specialized software like Autodesk Meshmixer [12] to work with the scanned data, meticulously processing and refining the model [13]. This software is particularly valued for its ability to handle complex polygonal meshes, enabling the designer to adjust the geometry for a perfect ergonomic fit, thereby ensuring that the prosthesis meets the functional and aesthetic needs of the patient. The designer's skill in manipulating these digital models is paramount in creating a detailed and accurate representation that can be seamlessly transitioned into the prototyping phase. Once the digital model is finalized, it is prepared for 3D printing, a process that requires careful optimization to ensure structural integrity and print quality. The use of additive manufacturing technologies, such as 3D printing, allows for the creation of a high-quality physical prototype that accurately replicates the digital design. This prototype undergoes further refinement and fitting, guided by medical expertise, to ensure that the prosthesis provides optimal comfort and functionality. The end result is a high-quality, custom-fit 3D prosthesis that significantly enhances patient mobility and quality of life. This approach is not limited to prosthetic limbs but extends to a wide array of healthcare applications, offering custom-fit solutions that cater to individual needs. A notable example is the 3D PrintAbility initiative by Nia Technologies, in collaboration with CBM Canada, The University of Toronto, and other partners, implemented at CoRSU Hospital in Uganda [14]. This project demonstrates how combining advanced 3D scanning and printing technologies with medical and design expertise can produce custom-fit prosthetic sockets and braces for children, significantly improving accessibility and quality of care in resource-limited settings. The high-quality outcomes of such initiatives underscore the transformative potential of integrating medical expertise with advanced 3D digital design and additive manufacturing technologies in healthcare.

EXAMPLE 3.3 AND 3.4 DIGITALIZATION OF ANCIENT SCULPTING MODELS AND 3D DESIGN USING PHOTOGRAMMETRY AND 3D PRINTING

The digitalization of ancient sculpting models through photogrammetry technology represents an advancement in the preservation and study of cultural heritage, enabling the detailed presentation and analysis of historic artifacts while ensuring their longevity. This process involves capturing high-resolution images of ancient sculptures and converting them into detailed 3D models using sophisticated software applications such as Autodesk ReCap, Blender, and MeshMixer (the software is described in detail in the current book). The significance of this technological approach is underscored by its application to notable artifacts from the ancient city of Varna, formerly known as Odessos, where the unique visual and functional attributes of these historical pieces are meticulously preserved and studied [15]. The place of research for ancient models is placed in the Archaeological Museum of Varna, Bulgaria [16]. The photogrammetric measurements and calculations facilitate the accurate reconstruction of these ancient models, providing a valuable resource for both scholarly research and public education. In particular, the digitization of an ancient stone angel model shows how 3D printing technology can be employed to recreate and redesign historic artifacts, albeit for personal, noncommercial purposes. This innovative process not only enhances the understanding and appreciation of ancient treasures but also serves as an educational tool, showcasing advanced computer techniques for teaching and learning. By offering a comprehensive guide to photogrammetry and its applications, this approach supports specialists in the digital field, including designers, engineers, and archaeologists, in their efforts to preserve and promote ancient cultural assets. This digital preservation and re-presentation initiative not only safeguards these invaluable artifacts but also fosters greater accessibility and engagement with the shared heritage. The digitalization of ancient models for research purposes involves a meticulous, multistep process designed to capture, analyze, and reproduce historical artifacts with high precision and clarity. Initially, the process begins with capturing detailed photographs of the artifact using a high-quality photo camera, ensuring that every angle and detail of the object is documented. These images are then processed through photogrammetry calculations using AutoDesk ReCAP software, which reconstructs the 3D structure of the artifact from the photographic data, effectively transforming 2D images into a detailed digital model. The resulting 3D model is imported into Blender software, where it is further refined and edited to enhance its accuracy and visual quality. This step involves manipulating the model to correct any distortions and to prepare it for the next phase. Following this, the refined model is imported into AutoDesk Meshmixer, where it is transformed into a solid object with precision, ensuring that the digital representation is both geometrically accurate and suitable for 3D printing. Finally, the solid model is used to create a physical replica through 3D printing, which provides a tangible version of the artifact that can be used for educational and noncommercial purposes. This process serves multiple research objectives: it explores and optimizes the capabilities and calculations of photogrammetry through modern technology, enhances the interoperability between different software systems and file formats, ensures the creation of high-quality 3D geometries, systematizes the transfer of data for 3D printing, and ultimately contributes to the redesign and preservation of historical artifacts for educational use [17].

Example 3.3 Digitalization of Ancient Sculpting Models and 3D Design Using Photogrammetry [15–17]

Real photo of the ancient model	Number of captured photos	3D digital model (with texture)	3D digital model (no texture)
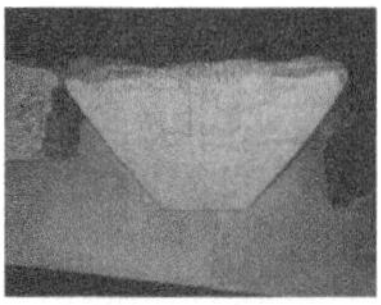	The photogrammetry result from 24 captured photos		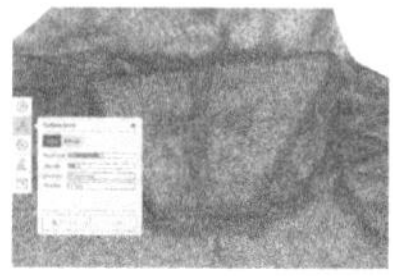
	The photogrammetry result from 32 captured photos		
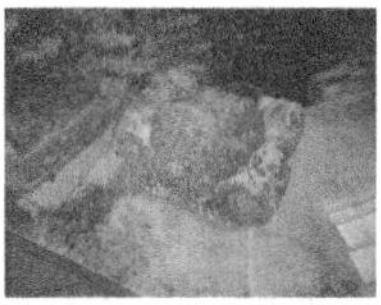	The photogrammetry result from 22 captured photos		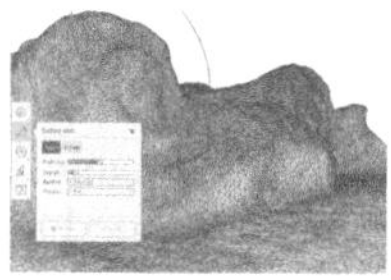
	The photogrammetry result from 21 captured photos	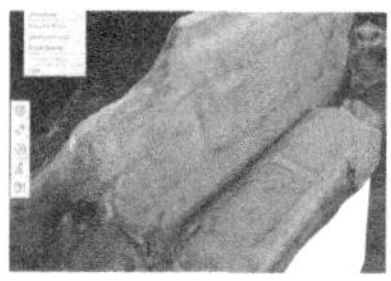	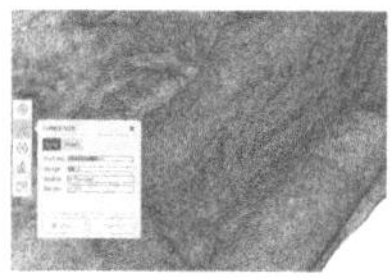

Example 3.4 Digitalization, Redesign of Ancient Sculpting Model of Angel Bas Relief and 3D Printing [16–19]

Real photo of the ancient model

The photogrammetry result from 22 captured photos

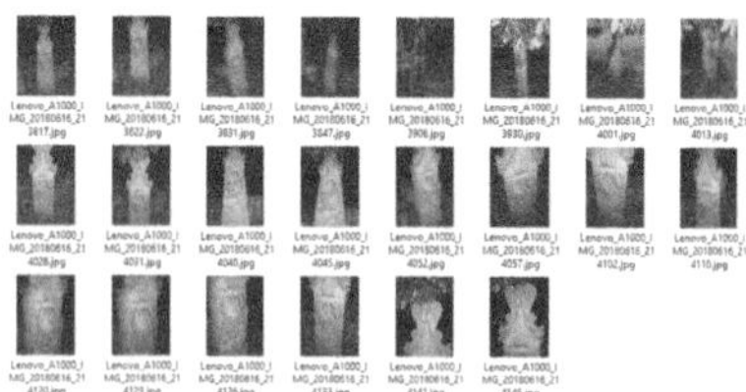

Autodesk ReCAP
The generated 3D model

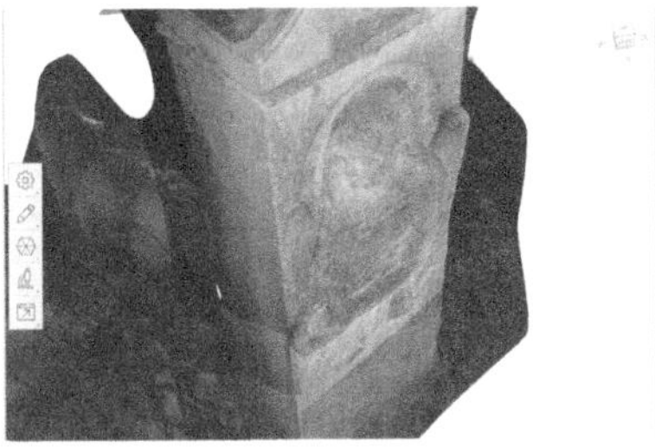

Autodesk ReCAP—The texture
The texture generated by ReCap

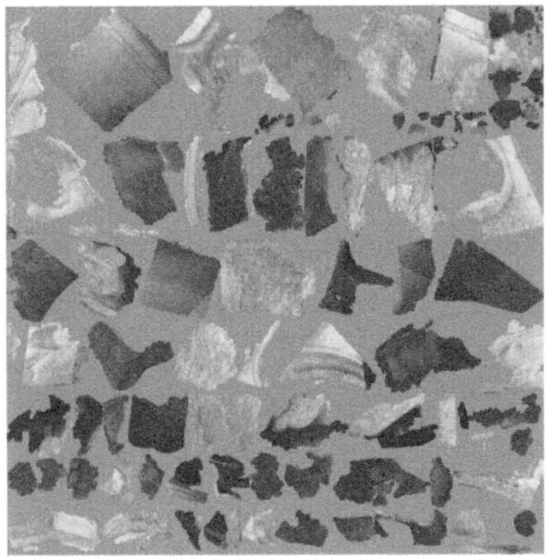

Autodesk ReCAP—slicing
The result & slicing

Autodesk ReCAP
The result & exporting to OBJ

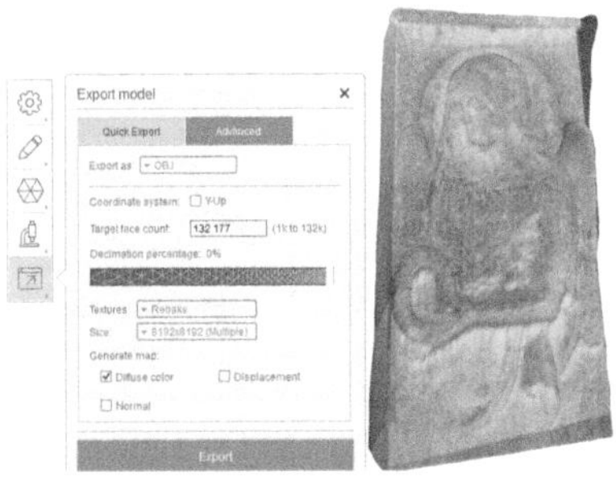

Blender—3D Free & Open Source Software.
Sculpting & redesign

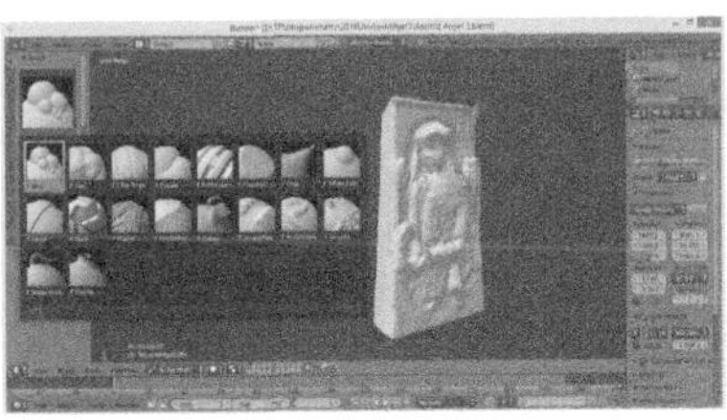

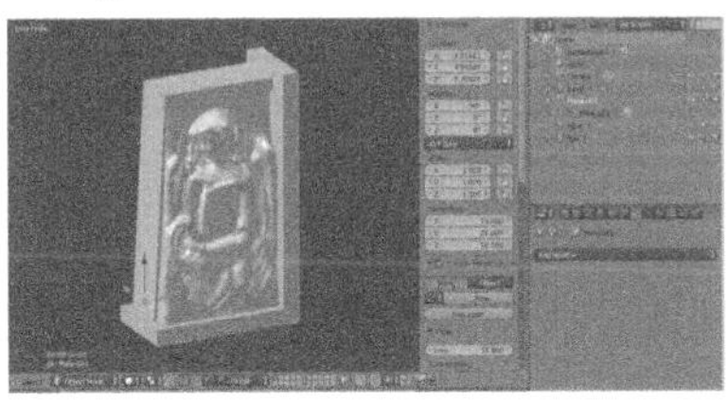

Blender—3D Free & Open Source
Software. Inside vision

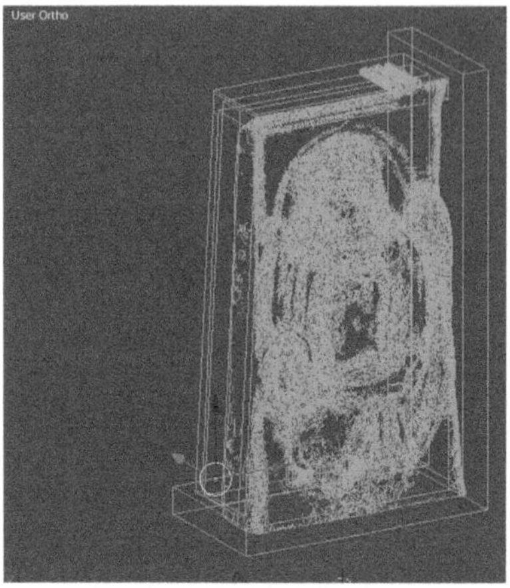

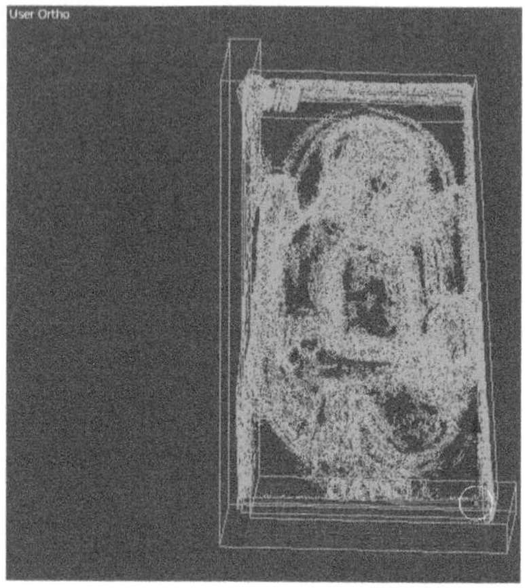

Blender software.Top & left views

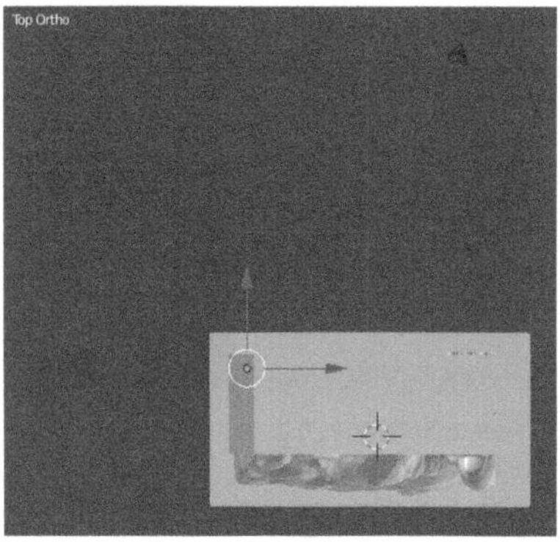

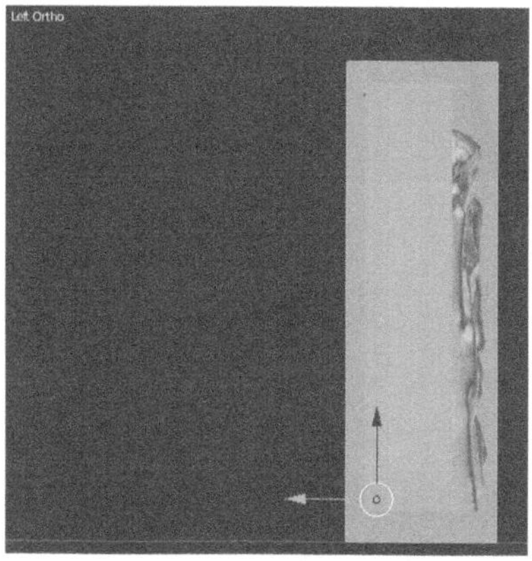

Blender software.
Exporting to OBJ / STL

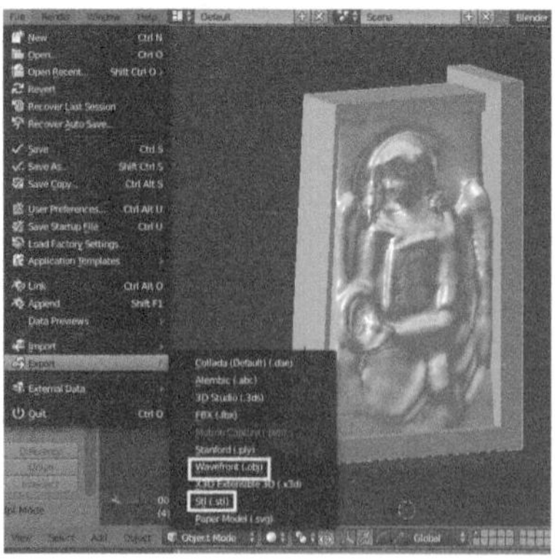

Autodesk MeshMixer
Making the solid model

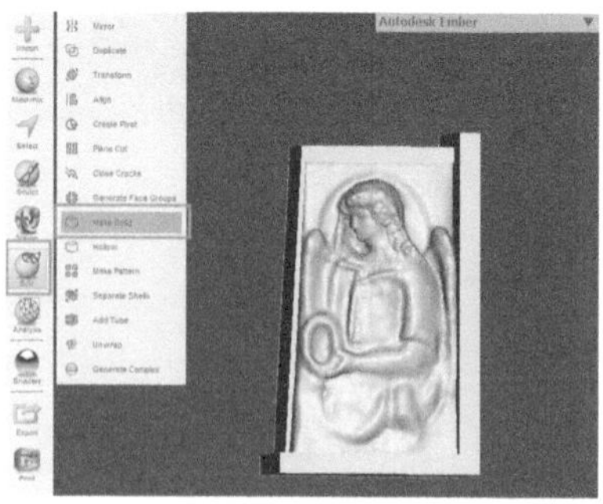

Autodesk MeshMixer
Making the solid model

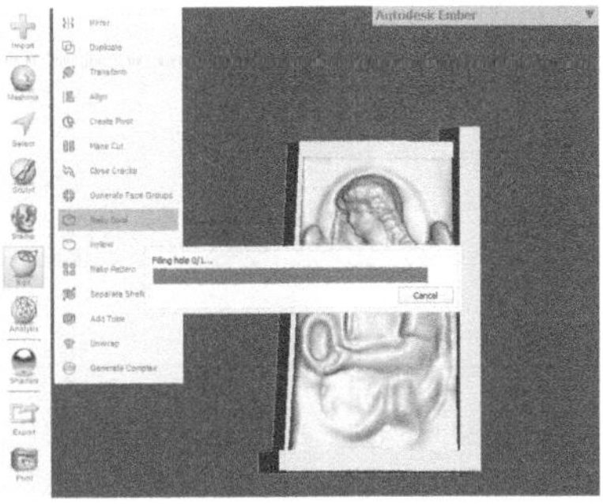

Blender preview. Zoom view

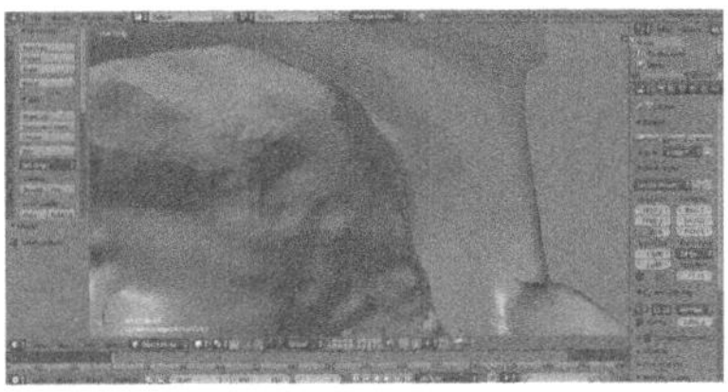

Blender preview

Blender preview. The mesh

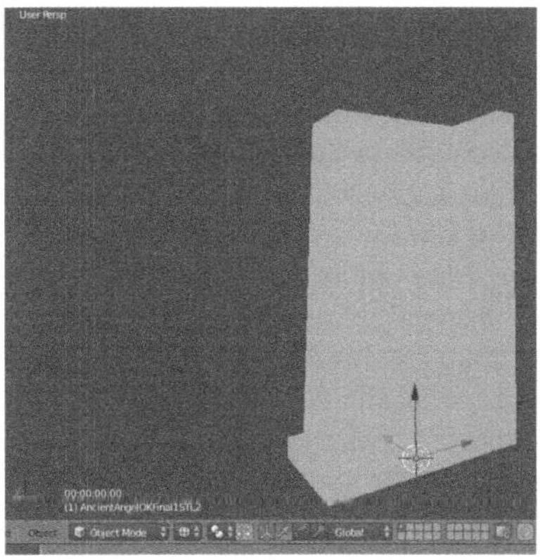

Texture [20]

Final Result: redesigned ancient angel + Textures. Blender Internal Render.

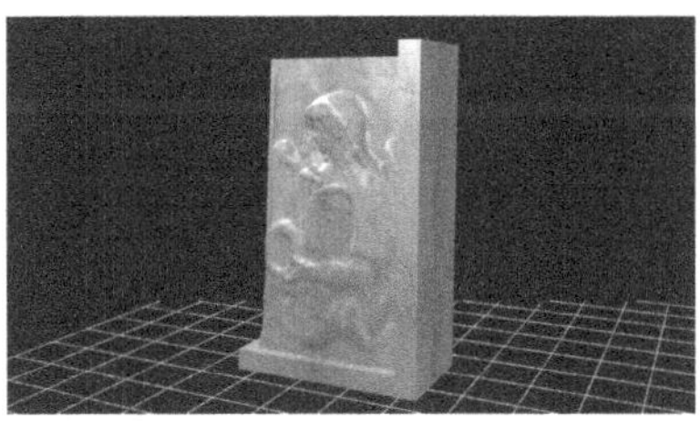

3D printing with Ultimaker 2 https://www.youtube.com/watch?v=5P3ytbzjp2U [21]

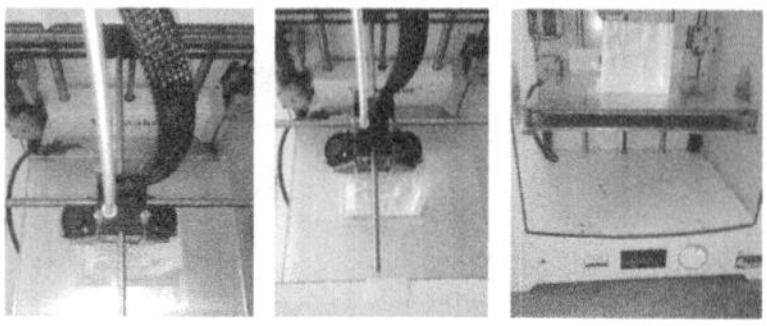

Final result: 3D printed redesigned ancient angel

REFERENCES

[1] XYZ hand scanner: 3D digitalization of a human model https://www.xyzprinting.com/en-US/product-level/PROFESSIONAL/3d-scanner-series
[2] XYZ hand scanner Pro: https://www.xyzprinting.com/en-US/product/3d-scanner-pro
[3] Creality CR-Scan Ferret Pro 3D (https://www.amazon.com/Creality-Scanner-Raptor-Printing-Handheld/dp/B0CLV1N3T6?th=1).
[4] AliceVision Meshroom (full free software) https://alicevision.org/
[5] 3DF Zephyr software https://www.3dflow.net/3df-zephyr-photogrammetry-software/
[6] Autodesk ReCap software https://www.autodesk.com/products/recap/overview?term=1-YEAR&tab=subscription&plc=RECAP
[7] RealityCapture software https://www.capturingreality.com/realitycapture
[8] SolidWorks: Stratasys printers https://www.solidworks.com/partner-product/stratasys-idea-series-3d-printers
[9] Stratasys printers: Mojo and uPrint SE™, uPrint SE Plus™, https://amtekcompany.com/doc/Idea%20Series/Idea%20Series%20brochure%20-%20English%20web.pdf
[10] SolidWorks: SolidView Stratasys Direct Manufacturing https://www.solidworks.com/partner-product/solidview
[11] Snapmaker https://www.snapmaker.com/about-us
[12] Autodesk Meshmixer software https://www.meshmixer.com
[13] Dovramadjiev, T., & Dimova, R. (2022). ERASMUS + Ergodesign Project Improving Digital Skills for Ergonomics AND Bioengineering Innovations for Inclusive Health Care TUV Software Description Autodesk Meshmixer. Zenodo. https://doi.org/10.5281/zenodo.13294250
[14] Nia Technologies' 3D PrintAbllity at CoRSU Hospital in Uganda https://www.youtube.com/watch?v=1gExokQ9Iys&t=282s
[15] Archaeology in Bulgaria. and Beyond. Odessos (Odessus) – Varna, Bulgaria https://archaeologyinbulgaria.com/background-infonotes/fortresses-cities/odessos-varna-bulgaria/
[16] Archaeological Museum - Varna, Bulgaria https://www.museumvarna.com/en/exposures/archaeological-museum/
[17] Dovramadjiev, T. (2024). Photogrammetry Methodology, *Digitalization of Sculpting Models & Design*. Zenodo. https://doi.org/10.5281/zenodo.13294270
[18] Dovramadjiev, T. (2019). Digitalization of Ancient Sculpting Models and 3D Design Using Photogrammetry Methodology. *International Journal of Engineering and Management Sciences*, 4(1), 458–465. https://doi.org/10.21791/IJEMS.2019.1.57

[19] Dovramadjiev, T. (2018). ISCAME 2018 Digitalization of Ancient Sculpting Models and 3d Design Using Photogrammetry Methodology. Zenodo. https://doi.org/10.5281/zenodo.13294685
[20] Pexels. Texture. https://www.pexels.com/photo/background-canvas-close-up-colour-1020320/; https://web.archive.org/web/20181004150950/https://www.pexels.com/photo/background-canvas-close-up-colour-1020320/
[21] 3D printing with Ultimaker 2. Re-Designed Ancient Angel - Stone Fragment - 3D Printing https://www.youtube.com/watch?v=5P3ytbzjp2U

4 3D Digital Design

The 3D design approach employs sophisticated three-dimensional digital modeling and simulation tools to craft highly accurate and detailed representations of physical objects and spaces, revolutionizing the way ergonomists and designers create and refine products by focusing on human interaction to enhance comfort, efficiency, and safety. This modern method addresses the limitations of traditional design, which often relies on generalized dimensions that may not cater to the diverse variability in human body sizes, shapes, and capabilities. By utilizing 3D digital design, designers can produce tailored solutions that precisely match the anthropometric data of specific user groups, thereby creating products and environments that are optimized for individual needs. For instance, office furniture, like chairs and desks, can be meticulously designed to accommodate a wide range of body types, thus mitigating the risk of musculoskeletal disorders and promoting better productivity. The process starts with crafting a detailed digital model using 3D polygonal-mesh software and computer-aided design (CAD) tools, enabling the model to be examined from any angle and modified in various ways [1–8]. This flexibility is particularly crucial in ergonomics, as it allows for a comprehensive evaluation of user interaction with products or spaces. Designers can simulate different scenarios to assess how factors such as reach, posture, and movement influence ergonomic outcomes, making adjustments to enhance functionality and comfort before physical prototypes are produced. 3D digital design includes dynamic simulations that model real-time human motion and interactions, offering insights into how design elements perform under actual use conditions. The 3D digital design integrates biometric and ergonomic data, such as body measurements, strength, and flexibility, into the design process to ensure that products and environments are adapted to users' physical capabilities [8–10]. This integration is particularly beneficial in designing medical devices, like surgical instruments or automotive interiors, ensuring that they fit user needs precisely, thus enhancing usability and safety. Beyond individual products, 3D digital design is invaluable for designing entire workspaces and environments. By creating detailed models of office layouts, factory floors, or public spaces, designers can assess how these environments support human activities and interactions holistically. This approach ensures that all aspects of the environment—furniture placement, lighting, and ventilation—are optimized to promote ergonomic principles and improve user well-being. For example, an ergonomic office design might feature adjustable desks, well-placed monitors, and appropriate lighting to minimize eye strain and support healthy postures, reflecting the comprehensive advantages of 3D digital design in creating spaces that genuinely cater to human needs [11–13].

Another aspect of 3D digital design in ergonomics and human factors is the development of assistive technologies and adaptive devices. These technologies are designed to support individuals with disabilities or specific physical needs, enabling them to perform tasks more easily and safely. Using 3D digital design, engineers can

DOI: 10.1201/9781003400929-4

create highly customized solutions that address the unique requirements of each user. For instance, a prosthetic limb can be designed to match the exact dimensions and functional needs of an amputee, improving comfort and functionality. Similarly, adaptive tools, such as ergonomic keyboards or mouse devices, can be tailored to reduce strain and enhance usability for individuals with repetitive strain injuries. 3D digital design supports the growing trend of user-centered design and participatory ergonomics. By involving end users in the design process, designers can gather valuable feedback and insights that inform the development of more effective ergonomic solutions. Virtual reality (VR) and augmented reality (AR) technologies, which are closely related to 3D digital design, provide immersive environments where users can interact with digital prototypes and provide real-time feedback. This collaborative approach ensures that the final design aligns with user needs and preferences, leading to higher satisfaction and better ergonomic outcomes [14–19].

In the context of education and training, 3D digital design offers powerful tools for teaching ergonomic principles and human factors concepts. Students and professionals can use digital models and simulations to explore the impact of design decisions on human performance and safety. This hands-on learning approach enhances understanding and retention of ergonomic concepts, preparing future designers and ergonomists to create user-friendly products and environments. 3D digital design tools can be used to develop ergonomics training programs that simulate real-world tasks and scenarios, allowing workers to practice and improve their skills in a controlled, risk-free environment [20–22].

In 3D modeling, meshes and solids are two fundamental types of models, each with its own characteristics and applications, crucial for fields like mechanical CAD. Solid models represent objects with volume and mass, defined by mathematical functions that describe their interior and exterior surfaces. They are volumetric, have properties like density, mass, and center of gravity, and allow for precise creation of fillets and rounds with exact radii. Solids support complex operations such as Boolean operations, making them ideal for engineering, manufacturing, and mechanical CAD, where precision and physical properties are essential. Mesh models, on the other hand, represent objects as a collection of vertices, edges, and faces, typically triangles or quadrilaterals, defining the object's shape. They are highly flexible, suitable for organic shapes, detailed surfaces, and textures, and are commonly used in gaming, animation, visual effects, and 3D printing. Meshes excel in applications involving deformation and animation but can be less precise and more labor-intensive for tasks like creating fillets. The key differences between these two types of models include fillet creation, where solids enable exact creation with specified radii while meshes require manual adjustment; accuracy, with solids offering higher precision suitable for strict dimensional tolerances; and model operations, with solids supporting complex operations like Boolean cuts with high precision, whereas meshes can be more complex and less reliable in such tasks. Understanding these differences is crucial for choosing the right tool and approach for a given task [23–29].

In Table 4.1 shown, the creation of a 3D solid model with a radius of 10 mm in Autodesk 123D Design software highlights the precision and accuracy inherent in solid modeling. In Autodesk 123D Design, users can specify exact dimensions, such as a radius of 10 mm, ensuring that the model meets precise engineering and

TABLE 4.1
Example of Creating a 3D Digital Sphere with Comparing Difference in 3D Geometry between Lead Free Software

Software	Autodesk 123D Design	Blender
Creating a 3D digital sphere	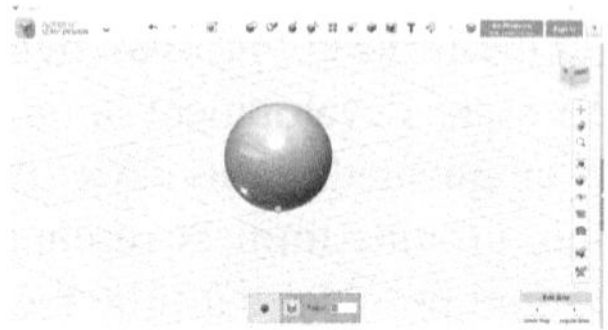	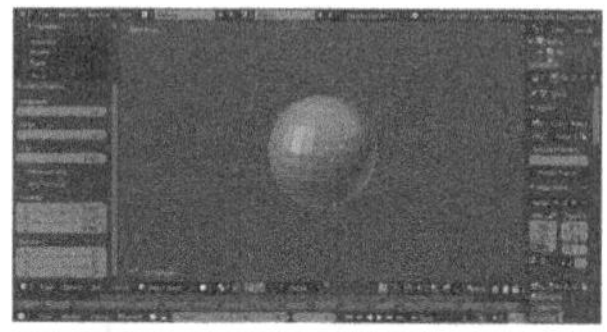
Editing the 3D digital sphere	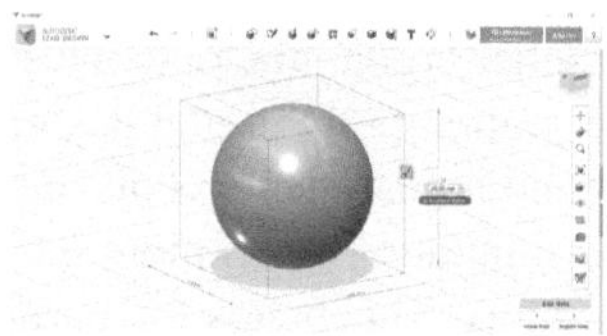	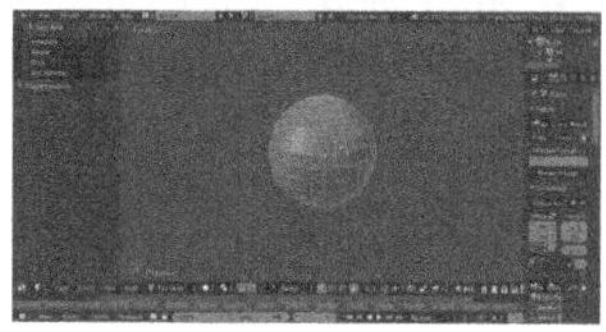

manufacturing standards. This level of accuracy is crucial for applications where strict tolerances and exact measurements are essential, such as in mechanical CAD and industrial design. The software's ability to handle complex operations like fillets and rounds with exact radii further underscores its suitability for detailed engineering work. For example, creating a smooth transition between surfaces (a fillet) can be done with precise radius values, which is critical in designing functional mechanical parts that must fit together perfectly and endure physical stresses.

In contrast, Blender3D, a popular open-source software known for its flexibility and robust toolset for animation and visual effects, approaches modeling differently. When creating a UV sphere in Blender, the user specifies the number of segments and rings (e.g., 32 segments and 16 rings) rather than exact dimensions like radius or diameter. This method is less precise in terms of exact measurements but offers significant advantages in terms of flexibility and ease of creating organic shapes. Polygonal-mesh modeling, as done in Blender, excels in scenarios where the overall form and surface detail are more important than exact dimensions. This makes Blender particularly well-suited for creating detailed biological models of humans, animals, and plants, where the complexity and texture of the model are prioritized over strict adherence to specific measurements.

The main difference between these two approaches lies in the precision and application focus. Autodesk 123D Design is engineered to provide the exact detail and accuracy required for mechanical and industrial applications. It allows users to input precise parameters and perform Boolean operations with high reliability, making it indispensable for tasks that demand rigorous standards and physical properties. On the other hand, Blender's polygonal-mesh modeling offers greater flexibility and creative freedom, enabling the creation of highly detailed and textured models that are ideal for animation, visual effects, and any application where the visual representation and organic complexity of the model are paramount.

Thus, while Autodesk 123D Design's ability to define exact radii and dimensions makes it a more precise tool for engineering and mechanical design, Blender's polygonal-mesh modeling capabilities provide unmatched flexibility for creating intricate and detailed organic shapes. Both tools serve distinct purposes and are invaluable in their respective fields, highlighting the importance of choosing the right software based on the specific needs and goals of the 3D modeling project.

The Table 4.2 depicts the creation of a 3D solid model of a cube with a one-edge fillet in Autodesk 123D Design software, showcasing the precision and accuracy inherent in solid modeling techniques. Unlike Blender3D, where a similar cube might be beveled with only a few segments to simulate a radius, Autodesk 123D Design uses mathematical definitions to accurately represent the volume and shape of objects. In polygonal-mesh software like Blender, achieving a visible "radius" involves increasing the number of segments significantly. Although this may create a visual representation of a curved edge, it's important to note that polygonal-mesh models do not possess inherent volume properties like solids do; they are composed of vertices and edges that approximate the surface of an object. The human eye typically cannot discern small differences between segment connections unless a very high number of segments is used, making polygonal-mesh modeling suitable for visual representation rather than precise engineering or manufacturing applications.

The mentioned software, Autodesk 123D Design and Blender, will be described in detail in the next chapters. These chapters will delve into the specific features,

TABLE 4.2
Example of Creating a 3D Digital Cube with Comparing Difference in 3D Geometry between Lead Free Software with Focus Radius (Fillet/Chamfer/Bevel)

	Autodesk 123D Design	Blender
Creating a 3D digital cube		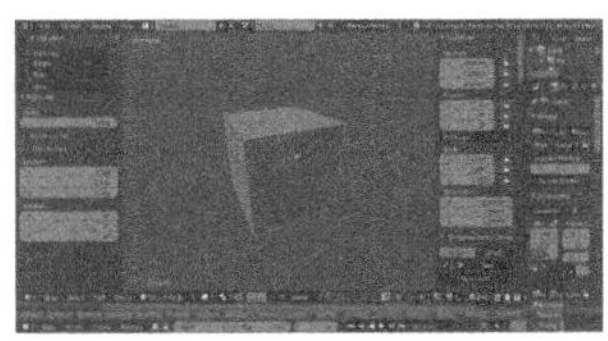
Editing a 3D digital cube. Autodesk 123D Design uses fillet for creating a radius, Blender software bevel option to simulate visible radius by subdivision parameters.	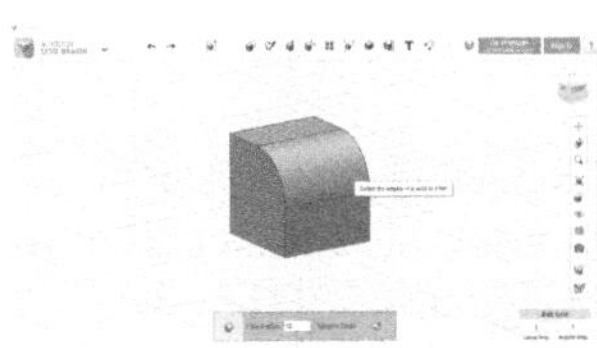	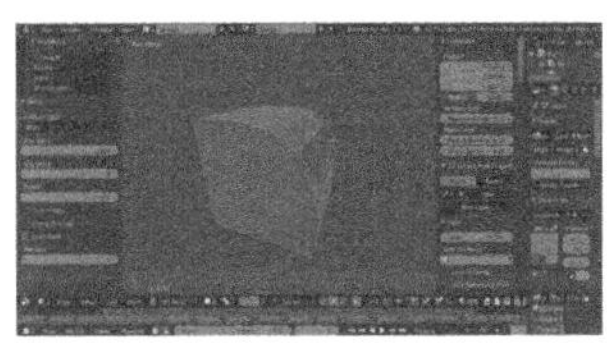
Fillet (Autodesk 123 Design software) vs Bevel (Blender software)	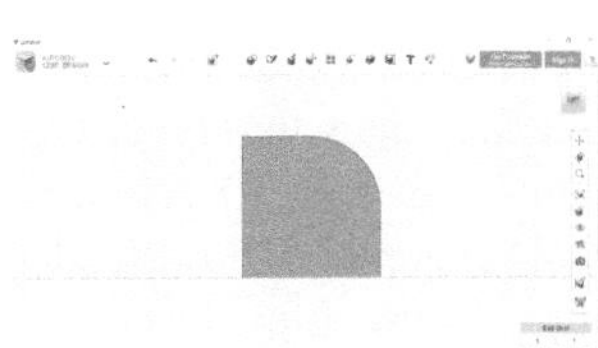	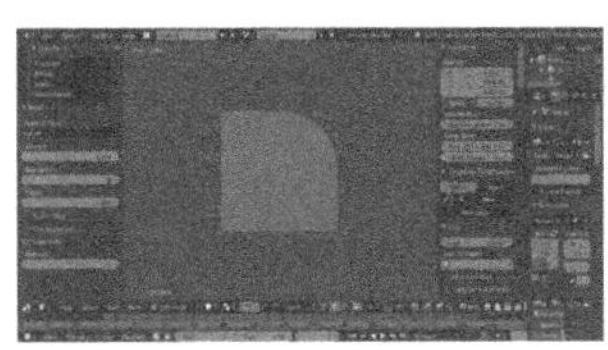

tools, and applications of each software, providing comprehensive insights into their capabilities and best uses.

The book prioritizes and emphasizes the use of free software that caters to a wide range of users, designers, and stakeholders, ensuring accessibility and empowerment in their design activities. By focusing on free software options, the book aims to democratize 3D digital design tools, making them available to a broader audience without financial barriers. These software solutions are chosen for their robust capabilities and user-friendly interfaces, which facilitate both basic and advanced design tasks. The book highlights how these free tools can be leveraged effectively across various disciplines, from mechanical engineering to artistic design, catering to diverse needs and skill levels. By promoting the use of accessible and powerful software, the book empowers readers to unleash their creativity, innovate freely, and contribute meaningfully to their respective fields without the constraints of expensive software licenses. In parallel with the global movement towards open-source design software, the author acknowledges and respects the contributions of commercial companies like Autodesk, Dassault Systèmes (SolidWorks), and other industry leaders who have developed incredibly powerful software over the last three decades. These companies have significantly advanced the field of 3D digital design with their sophisticated software suites, supported by a vast array of plugins and applications. The author recognizes the value brought by these commercial entities in pushing the boundaries of design capabilities, catering to professional users and industries that require robust features, technical support, and integrated solutions. By highlighting both open-source and commercial options with varying licensing models, the book aims to provide readers with a comprehensive understanding of the diverse tools available, empowering them to choose the most suitable software for their specific design needs and aspirations. For instance, 3D designers and users worldwide owe a great deal of gratitude to Autodesk for releasing powerful software such as 123D Design CAD and MeshMixer. These tools have democratized 3D design and rapid prototyping, making sophisticated design capabilities accessible to a broader audience. Similarly, SolidWorks has been instrumental with its retraining programs [30], effectively promoting its software globally and empowering designers across the world with accessible tools under various programs. These initiatives from Autodesk and SolidWorks have not only fostered innovation but also supported the growth of a vibrant community of designers, engineers, and enthusiasts who rely on these tools for creating everything from intricate prototypes to functional models and beyond.

AUTODESK MESHMIXER. One of the most significant software programs in 3D sculpting design is Autodesk MeshMixer. The Autodesk MeshMixer is a real bridge between the 3D digital and 3D physical design world, with free but closed-source software that has become indispensable for designers and experts specializing in 3D sculpting (Table 4.2). Renowned for its high-precision digital instruments, MeshMixer facilitates the direct creation of solids, making it an essential tool in the final stages of 3D geometry preparation. The software, which reached its final version (3.5; download: https://meshmixer.com/download.html and/or https://www.softpedia.com/get/Multimedia/Graphic/Graphic-Editors/MeshMixer.shtml) [31, 32] in April 2018 and ceased further development as announced in September 2021, is highly regarded for its robustness and completeness, needing little to no updates. Despite its discontinuation, it remains a cornerstone in the digital design process, particularly when used in tandem with other tools like Blender. This seamless integration supports a continuous cycle from reality to virtuality, crucial for fields such as bioengineering and ergonomic design, where real-world geometries often need to be digitized. The workflow typically involves conventional 3D software methods and advanced additive technologies like 2D and 3D scanning or photogrammetric capturing, ensuring that digital models accurately reflect real-world examples. To materialize these digital designs into physical objects through 3D printing or CNC machining, it is imperative to generate high-quality, solid 3D geometries that are free from any breaches or inconsistencies. Autodesk MeshMixer excels in this area, offering sophisticated sculpting tools and automated "mesh sewing" capabilities that ensure solid walls across the entire model, thereby facilitating the creation of structurally sound and manufacturable designs. The software's ability to "clean" meshes generated through photogrammetric digitization further enhances its utility, making it a powerful tool for precise and efficient 3D model preparation in ergonomics and human factors design.

Autodesk MeshMixer has the following main functions (https://www.meshmixer.com/) [Section 3, source 12]:

- Drag-and-Drop Mesh Mixing;
- 3D Sculpting and Surface Stamping;
- Robust Convert-to-Solid for 3D Printing;
- 3D Patterns & Lattices;
- Hollowing (with escape holes);
- Branching Support Structures for 3D Printing;
- Automatic Print Bed Orientation Optimization, Layout & Packing;
- Advanced selection tools including brushing, surface-lasso, and constraints;
- Remeshing and Mesh Simplification/Reducing;
- Mesh Smoothing and Free-Form Deformations;
- Hole Filling, Bridging, Boundary Zippering, and Auto-Repair;
- Plane Cuts, Mirroring, and Booleans;
- Extrusions, Offset Surfaces, and Project-to-Target-Surface;
- Interior Tubes & Channels;
- Precise 3D Positioning with Pivots;
- Automatic Alignment of Surfaces;
- 3D Measurements;
- Stability & Thickness Analysis;

Triangle meshes are fundamental structures in 3D graphics, utilized extensively by MeshMixer and other modeling tools. A triangle mesh is composed of three primary elements: vertices, edges, and triangles. Vertices are discrete points in three-dimensional space, which can also carry additional data like color or texture information. Edges are line segments that connect pairs of vertices, while triangles are formed by connecting three vertices, creating the basic building blocks of the mesh. In the context of MeshMixer, the term faces is often used interchangeably with triangles, though technically, faces can refer to any polygonal surface, not just triangles. Neighbors in a mesh refer to the relationships between these elements, such as triangles that share an edge, edges that share a vertex, or vertices that share an edge. An element is described as incident to another when it is a part of it; for instance, an edge is incident upon the triangles it connects. A mesh boundary consists of edges that are incident upon only one triangle, defining the edge of the mesh. If a mesh has such boundary edges, it is considered open, whereas a closed mesh has no boundary edges. A shell in a mesh refers to a connected set of triangles, meaning that you can traverse from any triangle to any other within the same shell without leaving the surface, also known as a connected component. Surface normals and tangents are essential for understanding the orientation of the mesh: the normal vector at a point is perpendicular to the tangent plane at that point, which fits parallel to the surface, with the normal vector pointing "straight out" of the surface. UV mapping involves assigning 2D coordinates (UVs) to each vertex, which allows for the application of textures to the 3D surface by "wrapping" a 2D image onto it. Manifold surfaces have edges that are shared by one or two triangles, ensuring that each edge connects to either one or two triangles. Non-manifold surfaces, on the other hand, can have edges shared by three or more triangles, which are less common and usually present only in specific scenarios. Understanding these concepts is crucial for working effectively with triangle meshes, especially in terms of mesh integrity, texturing, and geometric manipulation [33].

- Sculpt Tool [34]: Autodesk MeshMixer's Sculpt Tool offers a versatile and user-friendly environment for 3D sculpting, allowing artists and designers to create detailed surfaces even with lower poly counts, thanks to its dynamic remeshing infrastructure. The tool features two main types of brushes: volume and surface brushes, each serving different purposes and offering unique capabilities. The Volume brushes are typically used for broader, more structural changes, while the Surface brushes are ideal for finer, surface-level detail work. Users can easily switch between these brush types via the Brush panel or using hotkeys (2 for Volume and 3 for Surface). Each brush type includes multiple Brush Operations (BrushOps), such as the Zipper Brush, that can be selected from the Brushes flyout panel. Basic usage involves left-click-dragging to apply a brush operation, known as a "stamp," which can be spaced at intervals for the Surface brush or controlled continuously with the Flow parameter. Additional control is provided through strength and size parameters, adjustable via hotkeys, the mouse wheel, or the hotbox accessed by holding the spacebar. A key feature is the ability to access "opposite" effects by holding the Ctrl key, such as deflating instead of inflating. Users can also toggle between a primary and secondary brush operation within a single stroke using the Shift key, and they can maintain or adjust brush settings like Symmetry, Flow, and SymSnap. The tool also supports advanced features like a shared symmetry plane across tools, controlled through a 3D transformation widget, and a refinement panel that enables or disables dynamic remeshing to maintain mesh quality during sculpting. This panel includes settings for controlling edge length (Refine), minimum edge size (Reduce), smoothing (Smooth), and adaptive remeshing (Adaptivity), which helps preserve fine details while sculpting. The Adaptivity setting is particularly useful for preventing the loss of detail when working with larger brushes. Despite being less advanced than specialized tools like Mudbox or ZBrush, MeshMixer's sculpting tools offer a powerful and flexible environment for creating detailed 3D models, with the added advantage of not requiring extensive knowledge of subdivision levels or other technical details. This combination of accessibility and power makes MeshMixer an excellent choice for both beginners and experienced artists looking for a free, versatile sculpting tool.

- **Make Solid Tool** [35]: The Autodesk MeshMixer Make Solid tool is a sophisticated function designed to convert complex 3D models into watertight solids, which are essential for 3D printing and other applications requiring nonintersecting meshes. It operates through a detailed property panel with numerous settings, each influencing the tool's performance and output. Unlike some tools in MeshMixer, the Make Solid tool requires manual updates to the preview by clicking the Update button, as it does not automatically recompute previews with parameter changes; this can be time-consuming, and the Cancel button may not respond immediately, potentially delaying interruption of the process. Key parameters include solid accuracy and mesh density, which define the voxel grid's resolution, with cell size options providing precise control over the number of subdivisions; higher values result in more accurate results but at a greater computational cost and increased triangle count. The Offset Distance and Min Thickness sliders, available only in Accurate Solid Type mode, allow for the adjustment of shell offsets and the thickening of thin areas, respectively, although the latter can introduce artifacts due to the addition of small spheres. The Solid Type drop-down menu offers four modes: Fast, which re-projects vertices onto the input surface for a quick but potentially artifact-prone result; Accurate, which uses a signed distance field for better precision but at a higher computational cost; Block, which meshes voxel cubes for a distinctly blocky output; and Sharp Edge Preserve, which attempts to recover sharp edges post-reprojection, though this mode can be slow and imperfect. Color Transfer options enable the resampling of vertex colors or texture maps onto the solid mesh, with modes such as Vertex-to-Vertex and Material-to-Vertex adjusting how color data is applied, while No Colors mode disables this feature. The tool's advanced settings include Cull Edges Threshold, which removes small and thin triangles to clean up the mesh, though this can introduce artifacts; Close Open Boundaries, which fills holes in meshes but can lead to poor results with adjacent open patches if not disabled; Auto-Repair Result, which automatically fixes small errors in the mesh but can occasionally cause issues or crashes; and Transfer Groups, which attempts to map face groups from the input mesh to the output solid, though this can result in jagged boundaries and disconnected triangles, requiring further manual cleanup. Understanding these settings and their interactions is crucial for leveraging Make Solid effectively, as it approximates complex input shapes with voxel grids and generates solid models suitable for various applications while managing the trade-offs between computational effort, accuracy, and output quality.

- **File formats** [36]: Autodesk MeshMixer supports the following file formats: *.obj, *.ply, *.stl, *.amf, *.3mf, *.off, and *.mix. Autodesk MeshMixer supports a variety of import formats, catering to different 3D modeling needs. The STL (stereolithography) format, renowned as the standard for 3D printing, is unfortunately not ideal for mesh storage, though MeshMixer accommodates both its ASCII and binary versions. Binary STL files may include face colors, which MeshMixer converts into vertex colors, albeit with some potential blurring due to the ambiguity of color modes. The OBJ (Wavefront Object) format, widely utilized across industries, is handled with the most comprehensive support in MeshMixer. This format's arbitrary polygons are converted into triangles upon import, and while OBJ supports color-per-vertex extensions and texture maps (preferably in PNG or JPG formats), the handling of groups requires further detail as specified in the Preferences section. The AMF (Additive Manufacturing Format) is a newer alternative to STL designed for 3D printing, but MeshMixer's support is limited to triangle-mesh data, reflecting its preference for OBJ over AMF. PLY (Polygon File Format), often produced by 3D scanners, is supported in both ASCII and binary forms, and MeshMixer can read per-vertex colors from PLY files, though it notes the format's complexity and offers assistance for problematic files. OFF (Object File Format) is a simpler format that lists triangles, essentially a more basic version of OBJ, and is found in some older 3D databases. The 3MF (3D Manufacturing Format) support in MeshMixer is relatively new and may still present issues, but it does allow for vertex colors and texture/material data. Additionally, the MIX format allows for importing and appending .mix files to the current scene. It is important to note that MeshMixer discards any normals stored in these files, recalculating its own, and assumes millimeters for units unless specified otherwise, with the exception of formats like DAE and AMF that include unit information.
- **Print Tool** [37]: Autodesk MeshMixer offers a versatile print tool that facilitates the process of preparing 3D models for printing by integrating with third-party printing software rather than performing the printing directly within the application. To optimize the workflow, MeshMixer provides a range of functionalities centered around printer management and model preparation. It includes a feature for visualizing the print volume of selected printers, allowing users to see how their model fits within the printer's build area. This is accessed via the "Show Printer Bed" option in the View menu, which renders a wireframe box representing the printer's dimensions. The printer selection widget, visible in the upper right-hand corner of the MeshMixer window when "Show Printer Bed" is active, lets users select from a list of printers; those with associated software are displayed in bold for easy identification. Users can manage this list through the

"Printer Preferences" option, which opens the Preferences window where existing printers can be edited and deleted, or new ones can be added. This includes specifying the printer's name, print volume dimensions, and associated software. MeshMixer supports various units for volume dimensions but internally converts these to millimeters. Users can also link a specific software package to handle the actual printing process by providing details such as the software name, path to the executable, default file format (usually STL), and optional command-line arguments for the software. When a printer with associated software is selected and the "Print" button is clicked, MeshMixer exports a temporary mesh file and launches the designated software with this file as an argument. For instance, if the software path is \Applications\MyPrinter.exe and the command-line arguments specify slice height and file location, MeshMixer will format the command appropriately. It's important to note that some software, like MakerBot Print, does not accept mesh files from the command line, which is why older applications like MakerBot Desktop might be used by default. Mac users should be cautious about providing the full path to the executable rather than just selecting the .app bundle to ensure MeshMixer can properly interface with the software. If no associated software is available for the selected printer, the "Print" button will prompt users to open the printer's properties to add or adjust custom print software settings.

As Autodesk MeshMixer is used for its own design and goals or digital transfer of data from other software and additive technologies, the next software selected by importance for stakeholders is listed in separate subsections.

4.1 3D POLYGONAL-MESH MODELING

3D polygonal-mesh modeling is a fundamental technique in digital design, particularly well-suited for both sculpting and creating intricate biological models. At its core, this method revolves around the structure of interconnected polygons—most commonly triangles or quadrilaterals—referred to as faces. These faces are connected by edges and defined by vertices, which together form a mesh that accurately captures the surface geometry of a wide range of objects, environments, and human forms. This polygonal structure is crucial for optimizing ergonomic designs, as it allows designers to meticulously model complex surfaces with a high degree of precision. For example, in the design of ergonomic chairs or handheld tools, the careful adjustment of vertices and edges enables the fine-tuning of shapes to maximize comfort and usability. The ability to manipulate these elements makes the technique particularly effective for sculpting detailed, organic forms, such as human anatomy or other biological units, where capturing the subtleties of surface curvature is essential. The integration of anthropometric data into these models further enhances design precision by tailoring dimensions and proportions to the specific needs of the intended user population. This capability is invaluable in creating ergonomic products that not only fit but also support and enhance user comfort. Additionally, the detailed mesh structure allows for the creation of highly realistic virtual environments, which can be used for ergonomic assessments and simulations. These environments provide a platform for evaluating design choices in a virtual space before physical prototypes are built, saving time and resources while ensuring that the final product meets all ergonomic criteria [38–41].

4.2 3D SOLID MODELING (COMPUTER-AIDED DESIGN – CAD)

3D solid modeling through CAD marks a significant advancement in E/HF, improving how products and environments are designed to enhance human comfort, well-being, and efficiency. Unlike traditional surface modeling, CAD provides a volumetric, detailed view of objects, incorporating internal features essential for comprehensive ergonomic analysis. This industrial-standard approach enables designers to create precise models that integrate human anatomical and physiological characteristics, ensuring that products like office furniture, automotive interiors, and medical devices meet ergonomic standards. By incorporating anthropometric data, CAD tailors designs to accommodate diverse body sizes and shapes, essential for creating custom-fit solutions like orthotics and prosthetics. The ability to simulate user interactions and adjust dimensions digitally before production allows for the prevention of work-related musculoskeletal disorders and improves overall user comfort. CAD's role in virtual prototyping is particularly valuable, enabling multiple design iterations to be tested and refined without the need for costly physical prototypes. This capability is critical in optimizing workspaces and environments, such as industrial workstations, where CAD simulations can identify potential ergonomic issues and improve safety and efficiency. Additionally, CAD integrates advanced technologies like finite element analysis (FEA) and motion capture to simulate structural performance and human movement, resulting in products that are not only comfortable but also durable and safe. CAD enhances collaboration among interdisciplinary teams by providing detailed visual models that ensure ergonomic considerations are seamlessly integrated into the design process [42–45].

The integration of advanced CAD tools such as SolidWorks, FreeCAD, AutoCAD, Inventor, and Solid Edge into the field of 3D digital design in ergonomics and human factors represents a significant advancement in the ability to create, analyze, and optimize human-centered designs. These tools provide the necessary capabilities to ensure that products and environments are designed with a deep understanding of ergonomic principles, ultimately leading to improved user experiences. In detail, these CAD software support the design process as alphabetically listed:

- **AUTODESK 123D DESIGN**: Autodesk 123D Design is user-friendly yet powerful 3D modeling software ideal for both beginners and experienced designers. It enables users to create precise, manufacturable objects using intuitive tools that start with simple geometric shapes and build complex models. Key features include advanced Boolean tools for shape manipulation, integration with MakerBot 3D printers, customizable templates, and a comprehensive part library. The software offers robust design tools like rotate, scale, extrude, and stretch, alongside accurate measurement and mesh repair tools for 3D printing. Additionally, parametric modeling, real-time rendering, and simulation tools for motion and stress testing enhance design precision and performance, making it suitable for a wide range of applications from simple projects to complex engineering designs.

Features of Autodesk 123D Design [46]:

- ***3D Modeling:*** The core functionality of Autodesk 123D Design is its ability to create 3D models using simple geometric shapes. This makes it accessible for users to build complex models by combining basic shapes.
- ***Access Design Community*****:** Users can leverage the built-in community feature to share their ideas, seek inspiration, and collaborate with other designers. This community aspect fosters a collaborative environment where users can improve their designs and get feedback.
- ***Advanced Boolean Tools*****:** With advanced Boolean operations, users can combine, subtract, and intersect 3D shapes in sophisticated ways, enabling the creation of complex models from simpler components.
- ***Connect to MakerBot 3D Printers*****:** The application allows direct connection to MakerBot 3D printers, streamlining the process from design to physical creation. This seamless integration supports a more efficient workflow for those using MakerBot printers.
- **Customizable Templates**: Users can create custom parts and components, providing a personalized touch to their projects. This feature helps in creating repeatable designs and ensures consistency in multiple projects.
- ***Design Tools*****:** Autodesk 123D Design offers robust design tools that include rotate, scale, extrude, and stretch functionalities. These tools provide flexibility in manipulating shapes to achieve the desired design.
- ***Export 3D Designs*****:** Once a design is complete, users can export their 3D models, ready for 3D printing or further design work in other applications. This ensures compatibility with various 3D printers and other CAD software.
- ***Measurement Tools*****:** Accurate measurement tools are available to measure distances and angles between 3D shapes, ensuring precision in design and alignment.
- ***Mesh Repair Tools*****:** Before 3D printing, it is crucial to repair and prepare models. Autodesk 123D Design includes mesh repair tools to fix any issues that could affect the print quality.
- ***Parametric Modeling*****:** Users can create parametric designs that are essential for projects requiring repeatable and adjustable elements. This feature enhances the capability to make design modifications without starting from scratch.
- ***Part Library*****:** The application provides access to a library of ready-made parts, which users can incorporate into their designs, saving time and effort.
- ***Real-Time Rendering*****:** Real-time rendering allows users to see a visual representation of their 3D models as they work on them, providing immediate feedback on how the final product will look.
- ***Simulate 3D Printing*****:** Before printing, users can simulate the 3D printing process to check for potential errors, ensuring a smooth printing experience and reducing the risk of failed prints.

- ***Simulation Tools*:** Autodesk 123D Design includes simulation tools for motion and stress testing, allowing users to evaluate the performance and durability of their designs under various conditions.
- ***Subdivision Modeling*:** For more detailed and refined 3D models, subdivision modeling tools are available to add sub-surface details, enhancing the overall quality and intricacy of the designs.

The workplace of 123D Design software [47], includes: the Main Menu (left interface position) with conventional activities such as create a New file, Open file, Save file, Save a Copy, Import, Export as 3D, Export as 2D, 3D Print, Send To, Exit; Undo and Redo buttons; the Main Toolbar, which includes (from left to right): Undo, Redo, Transform, Primitives, Sketch, Construct, Modify, Pattern, Grouping, Combine, Measure, Text, Snap, and Material; Views (Interactive 3D View Cube, top/right interface position); a Navigation Bar (right interface position); Snap (Linear and Angular bottom/right interface position); a Grid; and the Graphics Area (Work and Digital Design Environment).

Autodesk 123D is available as freeware; the latest version, 2.2.14, is compatible with Windows 7, 8, 10, and 11 and is available for both 32-bit and 64-bit systems, providing a robust platform for diverse design needs [48].

- **AUTODESK AUTOCAD**: Autodesk AutoCAD is CAD software that equips architects, engineers, and construction professionals with precision tools to enhance their design processes. It allows users to design and annotate 2D geometry and 3D models with solids, surfaces, and mesh objects, providing a comprehensive suite for creating detailed and accurate representations. AutoCAD automates drafting tasks, enabling the placement of objects with AI, comparing drawings, creating schedules, publishing layouts, and more, significantly boosting efficiency and accuracy. Users can maximize productivity through customized workspaces and the integration of AutoLISP, APIs, and apps, tailoring the software to their specific needs and workflows. Autodesk AutoCAD's capabilities are very useful for developing ergonomic and human-centered designs, ensuring precision and functionality in various applications such as architecture, engineering, healthcare, and beyond. The ability to integrate precise measurements and automate complex design tasks makes AutoCAD indispensable for ergonomic analysis and human factor considerations, enabling the creation of environments and products that are both functional and comfortable for end users. The inclusion of AI-driven features in AutoCAD allows for advanced simulation and prediction of human interactions with designed spaces, which is essential for preemptive identification and mitigation of potential ergonomic issues. This aligns perfectly with the goals of the book, which emphasizes the importance of incorporating ergonomic principles in all stages of the design process to improve overall user experience and safety. AutoCAD serves as a prime example of how advanced software can be leveraged to meet the stringent demands of modern design and engineering, ensuring that human factors are seamlessly integrated into the digital modeling process [49]. AutoCAD provides a comprehensive suite of tools and functionalities that streamline the design and drafting process for architects, engineers, and designers alike. The interface features the Ribbon, which optimizes the workspace by consolidating tools and controls into a compact interface that can be positioned horizontally, vertically, or as a floating palette. The Menu Browser offers efficient navigation through classic drop-down menus, real-time searches, and quick access to recent documents and commands. Steering Wheels provide intuitive 2D and 3D navigation tools such as pan, zoom, and orbit directly from the drawing status bar. The InfoCenter facilitates information retrieval, displays product updates, and manages saved topics and RSS feeds. Tool Palettes allow for the organization and customization of tools from drawings or third-party developers. The Command Prompt executes commands swiftly, either by full name or alias, with Dynamic Input enabling tooltips for streamlined command entry. Viewports enable multiple views of a drawing with different styles, managed through commands or the View tab. The Status Bar provides essential information like cursor coordinates and quick access to drawing and display tools. ShowMotion accesses named views and animated sequences within the drawing, enhancing visualization capabilities. Lastly, the ViewCube offers visual orientation feedback and history navigation, while the Quick Access Toolbar stores frequently used commands for rapid execution, making AutoCAD a powerful tool for efficient and precise design workflows [50].

- **AUTODESK FUSION 360**: AUTODESK FUSION 360 is a product design and manufacturing software that integrates design, engineering, and manufacturing into a single platform, fostering collaboration and innovation in the design process. Fusion 360's comprehensive toolset enables users to create detailed 3D models, simulate real-world conditions, and generate toolpaths for CNC machining. It supports parametric, freeform, and mesh modeling, providing flexibility and precision essential for ergonomic and human factors considerations. The software also offers robust collaboration capabilities, allowing teams to work together in real time, share designs, and manage project data seamlessly, which is crucial for multidisciplinary ergonomic design projects. During the design process, particularly in the predesign testing phase, 3D digital design in ergonomics and human factors benefits significantly from Fusion 360's ability to simulate and analyze designs. This allows for the evaluation of human interaction with products, ensuring that designs meet ergonomic standards and user needs before physical prototypes are created. By connecting correct CAD data and facilitating predesign testing, Fusion 360 ensures that products are not only functional but also comfortable and safe for users. Fusion 360's compatibility with both Windows and Mac OS and its support for a wide variety of file formats ensure interoperability with other CAD software, making it accessible to a wide range of users focused on ergonomic design. In industries such as automotive, aerospace, and consumer products, Fusion 360 helps designers and engineers bring their ideas to life efficiently and accurately, which is essential in the field of 3D digital design in ergonomics and human factors. Autodesk Fusion connects the entire manufacturing process by integrating CAD, CAM, CAE, and PCB into a single solution, streamlining product development. This integration is particularly beneficial for ergonomic and human factors applications, enabling seamless teamwork in one workspace and eliminating repetitive tasks with automated workflows. By managing project data and facilitating collaboration, Fusion 360 supports the creation of designs that are optimized for human use, enhancing both efficiency and accuracy. Fusion 360 is a tool designed for collaborative product development, seamlessly integrating organic and solid modeling to facilitate manufacturable designs. All Fusion design data is securely stored in the cloud, accessible from anywhere, and supports collaborative group projects for controlled access. This platform connects the entire product development process on both Mac and PC, empowering users to explore and refine designs with sculpting, modeling, and generative design tools. With real-time cloud collaboration, teams can enhance productivity by iterating design ideas efficiently. Fusion 360 enables optimization and validation through assemblies, joint and motion studies, and simulations, culminating in photorealistic renderings and animations to communicate design intent. It supports rapid prototyping via 3D printing and CAM toolpath generation for CNC machining. Employing a top-down design methodology, Fusion 360 allows users to break down systems into subsystems for detailed refinement, ensuring that design intent is preserved from assembly down to component levels through features like joints, rigid groups, and contact sets. This approach enhances design flexibility and accuracy, making Fusion 360 an invaluable tool in modern product development workflows [51]. Fusion 360's capabilities allow for the modernizing of product development with tailored workflows and functionalities. This includes enhancing efficiency and accuracy through automated drawings and providing access to up-to-date progress information and controlled access to audit trails of changes, comments, and history. Fusion 360 enhances the ability to design and manufacture ergonomically optimized products efficiently, making it a strong tool for professionals in this field [52].

- **AUTODESK INVENTOR** is another powerful software in the CAD suite, particularly known for its mechanical design and simulation capabilities. Inventor provides advanced tools for creating detailed 3D models and conducting simulations to test the performance and ergonomics of designs. Its parametric modeling feature allows designers to easily modify designs based on ergonomic requirements, such as adjusting the height of a workbench or the curvature of a chair to improve user comfort and reduce strain [53].
- **FREECAD** offers a powerful, open-source alternative that democratizes access to advanced CAD capabilities. It supports a wide range of applications, including product design, mechanical engineering, and architecture. In the context of ergonomics, FreeCAD is particularly valuable for

educational purposes and small-scale projects, providing tools for creating and analyzing 3D models without the need for expensive software licenses. Its modular architecture allows users to customize and extend functionalities, making it adaptable to specific ergonomic design needs [54].
- **SOLID EDGE** offers comprehensive solutions for 3D product design, including synchronous technology that combines the speed and simplicity of direct modeling with the flexibility and control of parametric design. Solid Edge's robust simulation capabilities enable designers to evaluate the ergonomic aspects of products under real-world conditions, ensuring that designs not only meet functional criteria but also enhance user well-being. Its integrated analysis tools help in identifying potential ergonomic issues early in the design process, facilitating the creation of products that are both functional and user-friendly [55].
- **SOLIDWORKS** is renowned for its robust capabilities and user-friendly interface, making it a leading software in the field of 3D solid modeling. Its extensive suite of tools allows for the intricate design and simulation of ergonomic products, from office furniture to medical devices. SolidWorks facilitates the analysis of human interaction with these products through advanced simulation features, such as motion analysis and stress testing. This ensures that designs not only meet aesthetic and functional requirements but also adhere to ergonomic principles, enhancing user comfort and safety [56].

The CAD software allows for the detailed modeling of human anatomy and the simulation of human interaction with various products and environments. This capability is essential for designing ergonomic solutions that cater to the diverse needs of users. For instance, by using these tools, designers can create adjustable workstations that accommodate different body types and postures or develop medical devices that fit comfortably and safely on the human body.

REFERENCES

[1] Dovramadjiev, Tihomir (2015). Modern accessible application of the system Blender in 3D design practice. *Journal, XXV-th traditional International Scientific Conference—4-5-th June*, hotel "Maritsa Iztok"—Starozagorski Mineralni Bani, Volume, 5, ISSN 1314-4111 140, Pages 10 – 13, https://scholar.google.bg/citations?view_op=view_citation&hl=en&user=laiDnWYAAAAJ&cstart=20&pagesize=80&sortby=pubdate&citation_for_view=laiDnWYAAAAJ:ziOE8S1-AIUC

[2] Leroux, Frédéric, Mickaël Germain, Étienne Clabaut, Yacine Bouroubi, and Tony St-Pierre. 2024. "Improving Three-Dimensional Building Segmentation on Three-Dimensional City Models through Simulated Data and Contextual Analysis for Building Extraction" *ISPRS International Journal of Geo-Information* 13, no. 1: 20. https://doi.org/10.3390/ijgi13010020

[3] Wei, Chialing, Mohit Gupta, and Thomas Czerniawski. 2023. "Interoperability between Deep Neural Networks and 3D Architectural Modeling Software: Affordances of Detection and Segmentation" *Buildings* 13, no. 9: 2336. https://doi.org/10.3390/buildings13092336

[4] Župan, Robert, Adam Vinković, Rexhep Nikçi, and Bernarda Pinjatela. 2023. "Automatic 3D Building Model Generation from Airborne LiDAR Data and OpenStreetMap Using Procedural Modeling" *Information* 14, no. 7: 394. https://doi.org/10.3390/info14070394

[5] Kaźmierczak, Rafał, Robert Skowroński, Cezary Kowalczyk, and Grzegorz Grunwald. 2024. "Creating Interactive Scenes in 3D Educational Games: Using Narrative and Technology to Explore History and Culture" *Applied Sciences* 14, no. 11: 4795. https://doi.org/10.3390/app14114795

[6] Barszcz, Marcin, Jerzy Montusiewicz, Magdalena Paśnikowska-Łukaszuk, and Anna Sałamacha. 2021. "Comparative Analysis of Digital Models of Objects of Cultural Heritage Obtained by the "3D SLS" and "SfM" Methods" *Applied Sciences* 11, no. 12: 5321. https://doi.org/10.3390/app11125321

[7] Carbonell-Carrera, Carlos, Jose Luis Saorin, and Stephany Hess-Medler. 2020. "A Geospatial Thinking Multiyear Study" *Sustainability* 12, no. 11: 4586. https://doi.org/10.3390/su12114586

[8] Blümel, Florian Erwin, Mathias Schulz, Ralph Breithaupt, Norbert Jung, and Robert Lange. 2024. "Enhancing Resilience in Biometric Research: Generation of 3D Synthetic Face Data Using Advanced 3D Character Creation Techniques from High-Fidelity Video Games and Animation" *Sensors* 24, no. 9: 2750. https://doi.org/10.3390/s24092750

[9] Wyciślik, Łukasz, Przemysław Wylężek, and Alina Momot. 2024. "The Improved Biometric Identification of Keystroke Dynamics Based on Deep Learning Approaches" *Sensors* 24, no. 12: 3763. https://doi.org/10.3390/s24123763

[10] Samatas, Gerasimos G., and George A. Papakostas. 2022. "Biometrics: Going 3D" *Sensors* 22, no. 17: 6364. https://doi.org/10.3390/s22176364

[11] Minaoglou, Prodromos, Nikolaos Efkolidis, Athanasios Manavis, and Panagiotis Kyratsis. 2024. "A Review on Wearable Product Design and Applications" *Machines* 12, no. 1: 62. https://doi.org/10.3390/machines12010062

[12] Duplakova, Darina, Jan Duplak, and Rastislav Kascak. 2022. "Ergonomic Rationalization Sequence of Digital Lighting Design in the Working Environment" *International Journal of Environmental Research and Public Health* 19, no. 12: 7275. https://doi.org/10.3390/ijerph19127275

[13] Corsini, Lucia, and James Moultrie. 2019. "Design for Social Sustainability: Using Digital Fabrication in the Humanitarian and Development Sector" *Sustainability* 11, no. 13: 3562. https://doi.org/10.3390/su11133562

[14] AlGerafi, Mohammed A. M., Yueliang Zhou, Mohamed Oubibi, and Tommy Tanu Wijaya. 2023. "Unlocking the Potential: A Comprehensive Evaluation of Augmented Reality and Virtual Reality in Education" *Electronics* 12, no. 18: 3953. https://doi.org/10.3390/electronics12183953

[15] Tan, Yi, Wenyu Xu, Shenghan Li, and Keyu Chen. 2022. "Augmented and Virtual Reality (AR/VR) for Education and Training in the AEC Industry: A Systematic Review of Research and Applications" *Buildings* 12, no. 10: 1529. https://doi.org/10.3390/buildings12101529

[16] Yu, Rongrong, Ning Gu, Gun Lee, and Ayaz Khan. 2022. "A Systematic Review of Architectural Design Collaboration in Immersive Virtual Environments" *Designs* 6, no. 5: 93. https://doi.org/10.3390/designs6050093

[17] Park, Hyejin, and Seungyeon Choo. 2022. "A Development of Visualization Technology through AR-Based Design Checklist Connection" *Applied Sciences* 12, no. 12: 6126. https://doi.org/10.3390/app12126126

[18] Banerjee, Shakti, Anirban Chowdhury, and Nilakshi Yein. 2023. "User Experience Evaluation of a Virtual Reality Tool Used for 3D Modelling in Industrial Design Education: A Study in the Indian Context" *Designs* 7, no. 5: 105. https://doi.org/10.3390/designs7050105

[19] Wong, Jia Yi, Abu Bakr Azam, Qi Cao, Lihui Huang, Yuan Xie, Ingrid Winkler, and Yiyu Cai. 2024. "Evaluations of Virtual and Augmented Reality Technology-Enhanced Learning for Higher Education" *Electronics* 13, no. 8: 1549. https://doi.org/10.3390/electronics13081549
[20] Mendoza-Muñoz, Ismael, Mildrend Ivett Montoya-Reyes, Aidé Aracely Maldonado-Macías, Gabriela Jacobo-Galicia, and Olivia Yessenia Vargas-Bernal. 2024. "A Hierarchical Axiomatic Evaluation of Additive Manufacturing Equipment and the 3D Printing Process Based on Sustainability and Human Factors" *Processes* 12, no. 6: 1083. https://doi.org/10.3390/pr12061083
[21] Urquhart, Lewis, Andrew Wodehouse, Brian Loudon, and Craig Fingland. 2022. "The Application of Generative Algorithms in Human-Centered Product Development" *Applied Sciences* 12, no. 7: 3682. https://doi.org/10.3390/app12073682
[22] Boschetti, Giovanni, Maurizio Faccio, and Irene Granata. 2023. "Human-Centered Design for Productivity and Safety in Collaborative Robots Cells: A New Methodological Approach" *Electronics* 12, no. 1: 167. https://doi.org/10.3390/electronics12010167
[23] Autodesk MeshMixer: About Triangle Meshes https://help.autodesk.com/view/MSHMXR/2019/ENU/?guid=GUID-7B6A26A2-1E8A-4352-99A5-6C4026D5B89D
[24] Blender software. Structure https://docs.blender.org/manual/en/latest/modeling/meshes/structure.html
[25] SolidWorks: Basic Concepts https://help.solidworks.com/2025/english/SolidWorks/sldworks/c_basic_concepts_solidworks_fundamentals.htm?id=0fd75a05588540628b2b3f339ddc705b#Pg0
[26] Dovramadjiev, Tihomir (2019). Blender Tutorial. Creating Sweep Similar Like SolidWorks 3D CAD System. Zenodo. https://doi.org/10.5281/zenodo.13331706
[27] Dovramadjiev, Tihomir (2019). Blender Tutorial. Creating Loft Similar Like SolidWorks 3D CAD System Using Bridge Edge Loops. Zenodo. https://doi.org/10.5281/zenodo.13331889
[28] Dovramadjiev, Tihomir (2019). Blender Tutorial. Creating Loft Cut Similar Like SolidWorks 3D CAD System + Keys for Round Self Mesh. Zenodo. https://doi.org/10.5281/zenodo.13332008
[29] Dovramadjiev, Tihomir (2018). Blender software. Example 1. Simple 3D Models (Table + Chairs). Subdivide, Extrude & Bevel. Zenodo. https://doi.org/10.5281/zenodo.13332045
[30] SolidWorks: Retraining programs (https://www.controleng.com/articles/free-solidworks-3d-cad-software-training-part-of-engineering-stimulus-package/)
[31] Autodesk MeshMixer: Direct download https://meshmixer.com/download.html
[32] Autodesk MeshMixer: Softpedia download https://www.softpedia.com/get/Multimedia/Graphic/Graphic-Editors/MeshMixer.shtml)
[33] Autodesk MeshMixer: Help https://help.autodesk.com/view/MSHMXR/2019/ENU/?guid=GUID-7B6A26A2-1E8A-4352-99A5-6C4026D5B89D
[34] Autodesk MeshMixer: Sculpt Tool https://help.autodesk.com/view/MSHMXR/2019/ENU/?guid=GUID-6D77C4D0-0E73-411F-B00F-F71508D00DEA
[35] Autodesk MeshMixer: Make Solid Tool https://help.autodesk.com/view/MSHMXR/2019/ENU/?guid=GUID-F8C64203-AFB7-449C-A395-2C029A80BB6C
[36] Autodesk MeshMixer: File Formats https://help.autodesk.com/view/MSHMXR/2019/ENU/?guid=GUID-D11FDE6E-0279-4909-9B9C-1E115506ED9B
[37] Autodesk MeshMixer: Print Tool https://help.autodesk.com/view/MSHMXR/2019/ENU/?guid=GUID-5CE6EBE9-B1D8-4D2F-9C99-0C6CF1DF8CB3
[38] Münster, S. et al. (2024). 3D Modeling. In: *Handbook of Digital 3D Reconstruction of Historical Architecture*. Synthesis Lectures on Engineers, Technology, & Society, vol 28. Springer, Cham. https://doi.org/10.1007/978-3-031-43363-4_6

[39] Luis, A., S. Lizbeth, C. Rosario, M. Byron. (2023). 3D Modelling of Freedom Summit for Virtual Environments. In: Botto-Tobar, M., Gómez, O.S., Rosero Miranda, R., Díaz Cadena, A., Luna-Encalada, W. (eds) *Trends in Artificial Intelligence and Computer Engineering. ICAETT 2022.* Lecture Notes in Networks and Systems, vol 619. Springer, Cham. https://doi.org/10.1007/978-3-031-25942-5_43

[40] Yang, Fan, You Li, Mingliang Che, Shihua Wang, Yingli Wang, Jiyi Zhang, Xinliang Cao, and Chi Zhang. 2022. "The Polygonal 3D Layout Reconstruction of an Indoor Environment via Voxel-Based Room Segmentation and Space Partition" *ISPRS International Journal of Geo-Information* 11, no. 10: 530. https://doi.org/10.3390/ijgi11100530

[41] Zheng, Yang, Yu Jieyu Zhao Chen Tang Chen, and Yu Shushi. 2021. "3D Mesh Model Classification with a Capsule Network" *Algorithms* 14, no. 3: 99. https://doi.org/10.3390/a14030099

[42] Camba, J.D., N. Hartman, G.R. Bertoline (2023). Computer-Aided Design, Computer-Aided Engineering, and Visualization. In: Nof, S.Y. (eds) *Springer Handbook of Automation. Springer Handbooks.* Springer, Cham. https://doi.org/10.1007/978-3-030-96729-1_28

[43] Molenaar, J.M., D. Ingrassia (2024). Computer-Aided Design. In: *Mastering Digitally Controlled Machines.* Maker Innovations Series. Apress, Berkeley, CA. https://doi.org/10.1007/978-1-4842-9849-7_1

[44] Stark, R. (2022). Major Technology 1: Computer Aided Design—CAD. In: *Virtual Product Creation in Industry.* Springer, Berlin, Heidelberg. https://doi.org/10.1007/978-3-662-64301-3_7

[45] Shivegowda, M.D., P. Boonyasopon, S.M. Rangappa et al. A Review on Computer-Aided Design and Manufacturing Processes in Design and Architecture. *Arch Computat Methods Eng* 29, 3973–3980 (2022). https://doi.org/10.1007/s11831-022-09723-w

[46] LO4D: Autodesk 123D Design https://autodesk-123d-design.en.lo4d.com/windows

[47] CMCSS: introduction to 123D DESIGN https://www.cmcss.net/wp-content/uploads/2018/06/123d-design-manual.pdf

[48] Autodesk 123D Design: Softpedia download https://www.softpedia.com/get/Science-CAD/Autodesk-123D.shtml

[49] Autodesk: AutoCAD Overview https://www.autodesk.com/products/autocad/overview?term=1-YEAR&tab=subscription

[50] Autodesk: AutoCAD Interface https://download.autodesk.com/us/autocad/learningpath/interface_overview/Interface_Overview_ACAD.htm

[51] Autodesk: Fusion 360 https://help.autodesk.com/view/fusion360/ENU/?guid=GUID-1C665B4D-7BF7-4FDF-98B0-AA7EE12B5AC2

[52] Autodesk: Fusion 360 Overview https://www.autodesk.com/products/fusion-360/overview?term=1-YEAR&tab=subscription

[53] Autodesk: Inventor https://www.autodesk.com/products/inventor/overview?term=1-YEAR&tab=subscription

[54] FreeCAD https://www.freecad.org/

[55] Siemens: Solid Edge https://solidedge.siemens.com/en/

[56] Dassault Systèmes – SolidWorks Corporation https://www.solidworks.com/

5 Open-Source Software (OSS) Licenses

In the field of digital design and ergonomics, the adoption of Open-Source Software (OSS) licenses such as the **GNU General Public License (GPL)** is of the highest importance, significantly influencing the development, distribution, and collaborative enhancement of software. OSS licenses, particularly the GPL, are designed to uphold and protect the freedoms associated with software use, ensuring that users can freely share, modify, and distribute software while contributing to a thriving, open-source worldwide society. The GNU GPL, introduced by the Free Software Foundation (FSF), serves as a cornerstone for free software, focusing on safeguarding user freedoms rather than merely addressing cost. This license guarantees that users can use software for any purpose, share it, study its workings, and modify it [1]. Table 5.1 shows the notable OSS digital design software and their main application.

A notable example of the GPL in action is Blender, a popular open-source 3D creation suite. Blender is released under the GNU GPL, which guarantees users the freedom to use the software for any purpose, whether personal, educational, or commercial. Users can distribute Blender, modify it, and share their modifications with others. The Blender Foundation is committed to maintaining Blender as free software, and its source code is generally licensed under GNU GPL Version 2 or later. Some components, like the Blender Cycles rendering engine, utilize more permissive licenses such as Apache 2.0, but the core software remains under GPL. This licensing model ensures

TABLE 5.1
Open-Source Software and Notable Software

Open-Source Software (OSS)	2D	3D	Priority Software Design Implementation by Users in Practice For:
Blender	2D: Grease Pencil	3D: Polygonal-Mesh Design	3D Various Design, Other
FreeCAD	2D: Drawings	3D: Solid Design	Industrial, Mechanical, Product, Other
LibreCAD	2D: Drawings	—	Technical Documentation, Other
Inkscape	2D: Vector Graphics	—	2D Graphics and Drawings, Other
Gimp	2D: Raster Graphics	—	2D Art, Digital Painting, Other
Krita	2D: Raster Graphics	—	2D Art, Digital Painting, Other
MakeHuman & MPFB	—	3D: Human Parametric Design	Creating 3D Digital Human, Rigging, Other

DOI: 10.1201/9781003400929-5

that all enhancements and modifications to Blender remain open-source, supporting a collaborative development community that continuously improves the software [2, 3].

Similarly, FreeCAD, another influential OSS in digital design, is licensed under the GNU Lesser General Public License (LGPL) for its source code, and the Creative Commons Attribution 3.0 License (CC-BY-3.0) for its documentation and website. The LGPL allows FreeCAD to be used, modified, and redistributed freely while permitting integration with proprietary software under compatible licenses. This flexibility supports both open-source development and the use of FreeCAD in various commercial and institutional settings, without imposing restrictive obligations on users [4].

- **LibreCAD** is distributed under the GNU General Public License version 2 (GPLv2), which permits free use, redistribution, and modification of the software. This open-source license fosters community-driven development while ensuring that the software remains accessible to all. The software is provided without any warranties, including those concerning its merchantability or suitability for a particular purpose [5].
- **Inkscape**, a vector graphics editor, is governed by the GNU GPLv2. This license applies to Inkscape's software, including its packages and extensions, ensuring that the program remains free and open-source. Users retain ownership of the files they create with Inkscape, such as SVG or PNG files, and can choose their own licensing for these works. The GPL ensures that Inkscape's source code and any modifications made to it are shared with the community, reinforcing the collaborative nature of open-source development [6].
- **GIMP**, another prominent OSS, operates under the GNU GPL. GIMP facilitates tasks such as photo retouching and image composition while allowing users to freely distribute and modify the software. The GPL's provisions ensure that all derivative works and modifications are also shared under the same license, maintaining an open and collaborative development model [7].
- **Krita** is licensed under the GNU GPL version 3, permitting its use for various purposes, including commercial work, educational installations, and free distribution. Commercial redistribution is restricted due to trademark ownership, but inclusion in free download sites is allowed. The source code is available for modification, with the condition that any distributed modified versions must also be released under the GPL. Krita incorporates third-party libraries with permissive licenses, all of which are clearly documented in the source code [8].
- **MakeHuman** and **MPFB** are open-source projects with a split licensing model. Core assets are released under the Creative Commons CC0 license, allowing unrestricted use, modification, and distribution without the need for payment, copyright notices, or attribution. The source code for MPFB is licensed under the GPL, while MakeHuman's source code is under the AGPL. These licenses ensure that modifications to the code must be shared under the same terms if distributed, promoting transparency and collaboration in the development community [9, 10]. (Based on priority 3D, the software Blender and MakeHuman are described in detail in the next chapters.)

The impact of OSS licenses like the GPL on digital design, particularly in ergonomics and human factors, is profound. These licenses ensure that software remain accessible, modifiable, and distributable, fostering innovation and collaboration across diverse fields. By removing financial barriers and promoting transparency, OSS licenses enable researchers, developers, and practitioners to work together more effectively, driving advancements in software tools that support ergonomic design and human factors research. The collaborative nature of open-source projects, combined with the robust legal framework provided by licenses like the GPL, creates an environment where creativity and innovation can flourish, benefiting users and contributors alike [11–18].

REFERENCES

[1] GNU General Public License https://www.gnu.org/licenses/gpl-3.0.html
[2] Blender: License https://www.blender.org/about/license/
[3] Dovramadjiev, Tihomir (2018). Blender software. 2.79b Version. About, License & Development. Zenodo. https://doi.org/10.5281/zenodo.13334225
[4] FreeCAD: License https://wiki.freecad.org/License
[5] LibreCAD: License https://docs.librecad.org/en/latest/appx/licenses.html
[6] Inkscape: License https://inkscape.org/about/license/
[7] Gimp: License https://www.gimp.org/about/
[8] Krita: License https://krita.org/en/about/license/
[9] MakeHuman: License https://static.makehumancommunity.org/about/license.html
[10] MPFB https://static.makehumancommunity.org/mpfb.html
[11] Eichstädt, T., and S. Spieker (2024). Open Source. In: *52 Stunden Informatik*. Springer Vieweg, Wiesbaden. https://doi.org/10.1007/978-3-658-41838-0_31
[12] Rathee, S., and A. Chobe (2022). Open Source Licenses. In: *Getting Started with Open Source Technologies*. Apress, Berkeley, CA. https://doi.org/10.1007/978-1-4842-8127-7_3
[13] Kuna, P., T. Kozík, S. Kunová, M. Šebo (2018). Software Tools for Creating and Presenting Virtual 3D Models. In: Auer, M., Guralnick, D., Simonics, I. (eds) *Teaching and Learning in a Digital World. ICL 2017. Advances in Intelligent Systems and Computing*, vol 716. Springer, Cham. https://doi.org/10.1007/978-3-319-73204-6_3
[14] Klimt, Jonathan, Niklas Eiling, Felix Wege, Jonas Baude, and Antonello Monti. 2023. "The Role of Open-Source Software in the Energy Sector" *Energies* 16, no. 16: 5855. https://doi.org/10.3390/en16165855
[15] Lindman, Juho. 2014. "Similarities of Open Data and Open Source: Impacts on Business" *Journal of Theoretical and Applied Electronic Commerce Research* 9, no. 3: 59–70. https://doi.org/10.4067/S0718-18762014000300006
[16] Wilson, Michael L., and Vakhtang Tchantchaleishvili. 2013. "The Importance of Free and Open Source Software and Open Standards in Modern Scientific Publishing" *Publications* 1, no. 2: 49–55. https://doi.org/10.3390/publications1020049
[17] Papachristou, Evridiki, Panagiotis Kyratsis, and Nikolaos Bilalis. 2019. "A Comparative Study of Open-Source and Licensed CAD Software to Support Garment Development Learning" *Machines* 7, no. 2: 30. https://doi.org/10.3390/machines7020030
[18] Kyratsis, Panagiotis, Konstantinos Kakoulis, and Angelos P. Markopoulos. 2020. "Advances in CAD/CAM/CAE Technologies" *Machines* 8, no. 1: 13. https://doi.org/10.3390/machines8010013

6 Open-Source Software

Open-Source Software (OSS) stands out for its cost-effectiveness and accessibility compared to proprietary software, which often comes with expensive licenses. The typically low or no-cost nature of OSS tools lowers the entry barriers for educational institutions, small businesses, and researchers, facilitating wider adoption of sophisticated 3D modeling and simulation technologies in ergonomic assessments and design. This democratization of technology empowers more organizations to conduct in-depth ergonomic studies and design interventions. OSS promotes transparency and reproducibility in ergonomic research by making software algorithms and methodologies open for scrutiny and validation by the community. This openness enhances the reliability and trustworthiness of ergonomic studies, particularly in critical areas like medical device design and workplace ergonomics. Ethically, OSS aligns with the principles of open access and equitable knowledge sharing, promoting the distribution of technological advancements in ergonomics and human factors in a manner that respects diverse cultural and regional requirements. OSS in 3D digital design for ergonomics and human factors represents a transformative force that drives innovation, enhances collaboration, and broadens accessibility. By enabling customization, fostering community engagement, and ensuring transparency, OSS empowers stakeholders to advance ergonomic design practices and create safer, more efficient environments for users worldwide. As technology evolves, the role of OSS in ergonomics will likely expand, further pushing the boundaries of design innovation and maintaining ergonomic principles at the forefront [1–5].

6.1 2D DESIGN OPEN SOURCE SOFTWARE

The 2D Design OSS assumes a foundational role, crucial for the early stages of design where conceptualization and planning are paramount before advancing to sophisticated 3D modeling and simulation. OSS tools for 2D design, such as Inkscape, LibreCAD, GIMP, and Krita, are indispensable for creating initial sketches, technical drawings, and schematics that delineate essential ergonomic aspects like spatial relationships, proportions, and basic functionalities. These 2D designs connect to ergonomic considerations by visualizing how products and systems will interact with users, setting the stage for detailed 3D work. The cost-effectiveness and accessibility of OSS democratize the design process, making it accessible to educational institutions, small enterprises, and independent designers without the financial burdens imposed by proprietary software. This openness fosters a broad participation base, enabling diverse contributors to engage in iterative design processes and enhance ergonomic solutions across various domains. OSS also ensures seamless integration with 3D modeling software, allowing for a coherent transition from 2D blueprints to complex 3D models, which is vital for maintaining design consistency and facilitating multidisciplinary collaboration. The collaborative nature of OSS promotes a

DOI: 10.1201/9781003400929-6

global exchange of knowledge and best practices, accelerating ergonomic innovation and ensuring that designs meet user comfort, safety, and usability standards. Additionally, the transparency inherent in OSS, through open-source code and peer review, strengthens the validity and reliability of ergonomic design methodologies, particularly in critical areas like healthcare equipment and consumer products. As technology advances, OSS's role in ergonomics will continue to evolve, driving forward new developments and maintaining a focus on creating user-centric, adaptable designs in a rapidly digitalizing world [6–9].

6.2 3D DESIGN OPEN SOURCE SOFTWARE

OSS, like Blender, FreeCAD, and MakeHuman, is transforming 3D digital design for ergonomic applications by offering powerful tools that allow users to develop tailored solutions across industries such as healthcare, automotive, and workspace optimization. These platforms integrate seamlessly, enabling efficient transitions from concept to simulation and optimization without losing critical details, which is essential for iterative prototyping and testing of ergonomic principles. The open nature of these OSS platforms fosters a collaborative community of designers and researchers, driving continuous innovation and ensuring that the tools evolve to meet emerging challenges. Their flexibility and advanced features are crucial for creating precise 3D designs that prioritize user comfort and safety, while the accessibility of these tools democratizes high-level ergonomic research, making them invaluable for educational institutions, startups, and nonprofit organizations. The OSS not only supports innovation but also cultivates a culture of knowledge sharing and improvement, making it an essential resource for ergonomic design [10–13].

A notable example of OSS implementation is the open-source 3D Blender, which achieved a monumental milestone when the animated film "Flow," created entirely using this software, won the prestigious Oscar for Best Animated Feature Film in 2025. This historic achievement not only highlights Blender's capabilities as a professional-grade tool for animation but also underscores its growing influence in the film industry. Directed by Gints Zilbalodis, "Flow" showcases the power of Blender's versatile features—from modeling and rendering to animation and visual effects—proving that OSS can compete with industry giants. The success of "Flow" has inspired a new wave of creators to explore Blender's potential, further solidifying its reputation as a leading platform for 3D design and animation. This groundbreaking accomplishment is a testament to Blender's transformative impact on the world of digital creativity (https://www.reuters.com/lifestyle/flow-wins-best-animated-feature-film-oscar-2025-03-03/).

The book "3D Digital Design in Ergonomics and Human Factors" highly recommends Blender, especially versions from 2.7x to the latest 4.2, due to its exceptional balance of features, versatility, and being free software. Blender stands out for its comprehensive range of functions and enhanced options, catering to various aspects of 3D design, from modeling and rendering to animation and game development. The software is frequently updated, ensuring users always have access to the latest tools and improvements, which, in practice, makes Blender a platform with virtually unlimited possibilities. The global Blender community is rapidly expanding, with a growing number of modern users every day. This is reflected in the impressive volume of free add-ons released daily or within a few days, significantly extending Blender's capabilities. For the latest updates and add-ons, users can follow Blender news on websites such as BlenderNation (https://www.blendernation.com) and Blender Artists (https://blenderartists.org), among others. Blender software is a "digital center" for state-of-the-art designers, providing a versatile platform to freely realize their creative ideas. It seamlessly accommodates a spectrum of projects, from conventional and advanced design to specialized ergonomics and 3D digital human models. Blender's extensive feature set, enhanced by highly recommended parametric add-ons, allows for intricate and diverse development paths. This makes it an ideal tool for a wide range of applications, empowering designers to innovate without the constraints of expensive software, thereby fostering a vibrant community of creative professionals and enthusiasts.

- **Blender 2.76** stands out as the most optimized and feature-rich version of the Blender 2.7x series, free for download software (file: Additional Resources: Table 6.1) tailored to run efficiently on low-power and older hardware. This makes it an invaluable tool for individuals and students who lack access to high-end computers. Despite its minimal system requirements, Blender 2.76 integrates the best tools and enhancements from the 2.7x family, including advanced rendering, improved compatibility, and significant user interface upgrades. These enhancements ensure smooth performance even on modest setups, making high-quality 3D modeling, animation, and rendering accessible to a broader audience. The inclusion of features like efficient viewport performance, streamlined modeling tools, and versatile add-ons ensures that users can engage in complex projects without needing cutting-edge technology. By balancing powerful functionality with low resource demands, Blender 2.76 democratizes digital content creation, providing a robust platform for learning and creativity regardless of hardware limitations. Blender 2.76 introduced significant enhancements across various areas, most notably in Cycles rendering and GPU compatibility. The new Point Density Texture generates voxel textures during shader compilation to minimize render time, although it is not yet optimized for memory efficiency, with future improvements planned through OpenVDB integration. AMD GPU stability and compatibility received a boost, particularly on Windows, Linux, and OSX El Capitan, expanding the range of usable GPUs. Camera zoom motion blur was implemented, with further refinements anticipated. The user interface saw major performance improvements, especially in the viewport, benefiting complex scenes, sculpting, and painting. Node Editor efficiency was increased with auto-offset functionality, and the File Browser was revamped for quicker, more intuitive navigation. Modeling updates included new tools like Flatten Faces and Edge Offset, better data transfer between meshes, and absolute grid snapping. Opensubdiv integration improved viewport playback performance, and the Sequencer gained a new Text effect type with subtitle export capabilities. Freestyle NPR rendering became more memory-efficient, and sculpting tools were enhanced with tiling options for seamless patterns. Animation saw the addition of Bone Selection Sets and a Clean Channels Tool, while the Game Engine featured a more polished Game Publishing Addon and other functional updates. Add-ons for FBX and OBJ import/export were refined, and new features were added to the UV Addon and Python API. The release also improved the animation/movie player, sculpting, datablock previews, and multiple link/append functionalities, along with numerous bug fixes and optimizations [14, 15].

- **Blender 2.79b** stands out as one of the best and most stable version in the Blender 2.7x series, improving upon versions 2.76, 2.77, and 2.78. While newer versions like Blender 4.2 offer advanced features and a revamped interface, many first and second-generation Blender developers consider 2.79b to be the pinnacle of the software. This version is renowned for its speed and efficiency, requiring minimal hardware resources and featuring a fast, noise-free internal render engine that excels in both graphical and animation rendering. Blender 2.79b is highly adaptable, available in both installable and portable versions, and is compatible with a wide range of operating systems. It supports numerous add-ons, enhancing its functionality and making it a versatile tool for various design needs. Despite the advancements in later versions, many platforms and developers continue to base their work on Blender 2.79b, appreciating its reliability, performance, and extensive community support. Blender 2.79b is the concluded version of Blender 2.7x series (not only because of the version number but also because it completes the interface design vision of the Blender conception). Blender 2.79b offers a comprehensive suite of improvements and new features that make it an essential tool for both beginners and seasoned professionals. This version introduces Denoise, which effectively eliminates rendering noise while preserving visual detail, ensuring smooth results without extending render times. The new Filmic Color Management system enhances photorealism and handles high dynamic range imagery more effectively. The introduction of the PBR Shader, based on Disney's principled BSDF, simplifies complex surface shading by combining multiple layers into a single, easy-to-use node. OpenCL rendering has been significantly optimized, providing much faster performance on AMD hardware and feature parity with NVIDIA CUDA, including support for Subsurface Scattering, Volumetrics, and optimized transparent shadows. The Shadow Catcher feature in Cycles allows for seamless integration of CGI elements with real-life footage by making objects receive shadows in a way that they can be easily composed onto another image. Cycles also benefits from multithreaded shader compilation, a new light sampling threshold, and speed improvements up to 20% for CPUs that support AVX2. Modeling and modifier enhancements include the new Surface Deform modifier for transferring motion from one mesh to another, multithreaded support for the Displace Modifier, and a new tool for setting custom normals from selected faces. Animation improvements feature a pinning option in the Graph Editor, color-coded channels in the DopeSheet, customizable motion path colors, and enhanced undo functionality. The Grease Pencil toolset is enriched with new interpolation tools, per-layer onion skinning, and various UI improvements, making it more versatile for 2D animation within a 3D environment. The user interface is more intuitive, with automatic scaling for high DPI displays, reorganized sequencer and UV editor panels, custom shortcuts, and improved tooltips [16, 17]. Alembic support has been significantly improved, enhancing compatibility and stability with new features like export of linked dupli-groups and sub-frame sampling. The Python API sees new capabilities, such as custom properties pointing to data blocks, custom render passes, and the ability to insert custom items in right-click menus. Blender 2.79b also includes a wide array of new add-ons that greatly expand its functionality, from parametric architectural design tools to sophisticated sky generation and advanced rigging systems for animation. This version is highly optimized for performance on modest hardware, requiring minimal resources while delivering professional-grade results. The internal render engine is fast and efficient, producing noise-free graphics and animations quickly. Its adaptability across various operating systems, along with the availability of both installable and portable versions, makes Blender 2.79b a versatile tool for designers, artists, and developers [18]. Blender 2.79b offers a highly flexible and user-friendly design, allowing users to customize their interface extensively. Users can easily adjust the size of interface elements and units (Table 6.2) [19–24] to suit their preferences, enhancing comfort and efficiency. The ability to select or create a personalized theme, in addition to choosing from base themes, provides further customization

TABLE 6.2
Blender 2.79b: Software Interface and Personal Design [19–24]

Blender 2.79 User Preferences Examples of Personal Design		
	User preference from default (left image) to Display Scale 1.5 (right image)	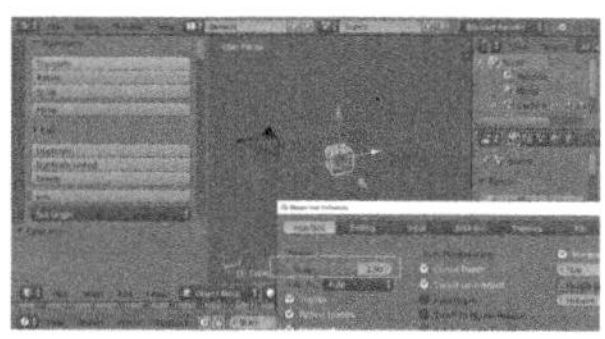
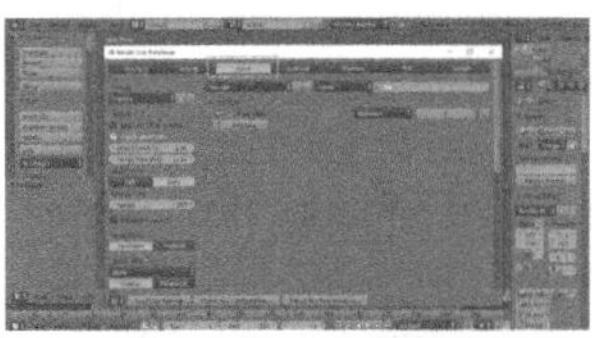	Example: Changing Export FBX (key "F"—left image) to (combination of "Alt + F"—right image)	
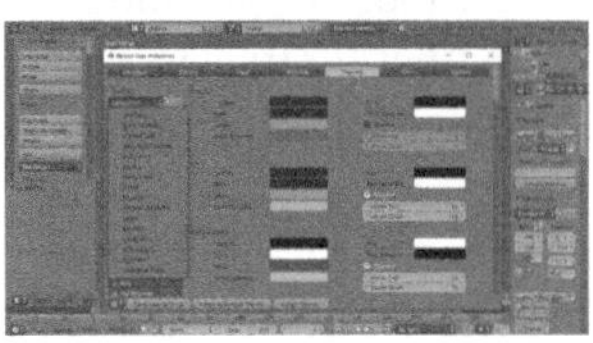	Change Interface Design (from Default—left image) to 24XBlues (right image)	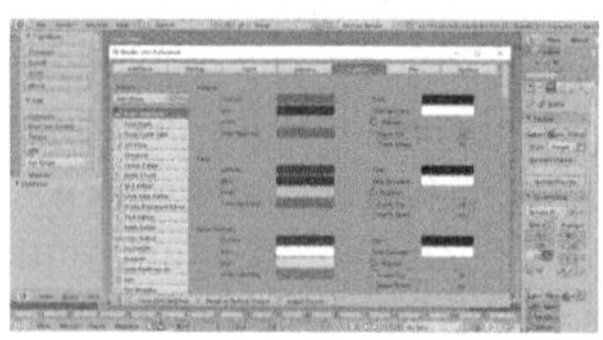

options. Blender 2.79b allows users to modify the input keyboard shortcuts according to their personal workflow or to learn from existing configurations. All these adjustments can be made through the User Preferences panel, enabling users to tailor Blender's environment to their specific needs and preferences, thereby optimizing their overall experience. The software is 100% free for download for OS: Windows, Linux, MacOS (file: Additional Resources: Table 6.3).

Blender OSS, developed by Dutch art director and software developer Ton Roosendaal [26], is widely recognized as the leading OSS for 3D content creation. Its comprehensive suite of tools supports every stage of the 3D design, including modeling, sculpting, texturing, rigging, animation, rendering, post-processing, and other. Blender's versatility and powerful capabilities have made it a cornerstone for both professional and hobbyist 3D designers, engineers, artists, and stakeholder. In parallel of Blender worldwide rising popularity, a developer Reiner "Tiles" Prokein on June 2, 2015 released **OSS Bforartists** [27], which aims to enhance Blender's user experience with a focus on graphical user interface (GUI) improvements and usability. Bforartists reimagines Blender's hotkey-centric workflow by introducing a more intuitive, icon-based interface, making it more accessible to new users and those coming from other software. This includes a streamlined keymap, extended and rearranged UI elements, a configurable toolbar with icon buttons, better tooltips, and improved default settings. Bforartists also maintains full compatibility with Blender files and addons, ensuring seamless transitions between the two platforms. The Bforartists project, driven by a federation of independent developers and a non-profit ethos, emphasizes ease of use and accessibility, particularly for hobbyists and indie developers. This dual development approach highlights the unique strengths of both Blender and Bforartists, with Blender providing a robust, professional-grade toolset, and Bforartists offering a user-friendly alternative that simplifies the learning curve and enhances productivity. Both platforms are testaments to the dedication of their respective communities, continually evolving to meet the diverse needs of the 3D design. Figure 6.1 shows an interface of **Bforartists** OSS version 1 which is synchronized wi Blender 2.79. The OSS software can be download from file links: Additional Resources: Table 6.4 (Bforartists version 1), Table 6.5 (Blender 4.2), and Table 6.6 (Bforartists version 4).

Figure 6.2 shows the interfaces of Blender 4.2 and Bforartists version 4.

6.3 INTERACTION BETWEEN 2D AND 3D OPEN SOURCE SOFTWARE

The interaction between 2D and 3D OSS sometimes is in correlation for creating designs that enhance user experience, safety, and efficiency across diverse applications. Initially, 2D OSS tools such as vector graphics editors like Inkscape and raster-based tools like GIMP are indispensable for the early stages of ergonomic design. These applications allow designers and engineers to generate detailed sketches, diagrams, and layout plans that form the blueprint for ergonomic solutions. Once these

FIGURE 6.1 Interface of Bforartists version 1. The software impressed with graphical icon design.

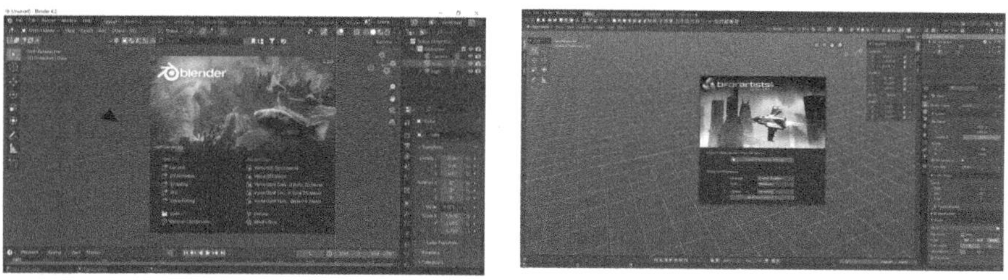

FIGURE 6.2 Interface of Blender 4.2 (left) and Bforartists version 4 (right).

foundational 2D designs are established, their integration with 3D OSS tools—such as Blender, FreeCAD, and OpenSCAD—enables a smooth transition from conceptual 2D visuals to intricate 3D models. This transition allows for the accurate simulation of physical environments, product interfaces, and user interactions, which is crucial for refining ergonomic designs. The synergy between 2D and 3D OSS fosters an iterative design process, essential for optimizing ergonomic solutions through continuous refinement based on user feedback, ergonomic principles, and performance evaluations. For example, designers can develop detailed 3D models of ergonomic furniture, workplace layouts, or interactive digital interfaces, incorporating considerations such as anthropometrics, cognitive ergonomics, and organizational ergonomics into their designs. This interaction also enhances visualization capabilities, enabling the creation of photorealistic renderings and animations that aid in communicating designs to interdisciplinary teams and stakeholders. Such visual representations are vital for demonstrating the ergonomic benefits and functionalities of designs, thereby facilitating decision-making and stakeholder consensus. The combined use of 2D and 3D OSS supports innovation and customization in ergonomic solutions, allowing designers to explore various design iterations and adapt principles to specific user needs and constraints. This flexibility is particularly valuable in fast-paced industries like healthcare, automotive design, and consumer electronics, where rapid prototyping and iterative refinement are crucial. Ethically and economically, the interaction between 2D and 3D OSS promotes sustainability by reducing material waste and resource consumption through virtual simulations and digital prototyping, leading to cost savings and environmental benefits. The open-source nature of these tools fosters knowledge sharing, collaboration, and community-driven innovation, advancing best practices and standards within the ergonomic design community [25–30].

REFERENCES

[1] Chronis, Christos, and Iraklis Varlamis. 2022. "FOSSBot: An Open Source and Open Design Educational Robot" *Electronics* 11, no. 16: 2606. https://doi.org/10.3390/electronics11162606

[2] Oberloier, Shane, and Joshua M. Pearce. 2018. "General Design Procedure for Free and Open-Source Hardware for Scientific Equipment" *Designs* 2, no. 1: 2. https://doi.org/10.3390/designs2010002

[3] So, Anita, Jacob M. Reeves, and Joshua M. Pearce. 2023. "Open-Source Designs for Distributed Manufacturing of Low-Cost Customized Walkers" *Inventions* 8, no. 3: 79. https://doi.org/10.3390/inventions8030079

[4] Leão, Celina P., Vinícius Silva, and Susana Costa. 2024. "Exploring the Intersection of Ergonomics, Design Thinking, and AI/ML in Design Innovation" *Applied System Innovation* 7, no. 4: 65. https://doi.org/10.3390/asi7040065

[5] da Silva, Adailton Gonçalves, Marcus Vinicius Mendes Gomes, and Ingrid Winkler. 2022. "Virtual Reality and Digital Human Modeling for Ergonomic Assessment in Industrial Product Development: A Patent and Literature Review" *Applied Sciences* 12, no. 3: 1084. https://doi.org/10.3390/app12031084

[6] Segura, J. 2022. Development of Computer Skills to Draw in the LibreCad from Virtual Learning Environments. In: Auer, M.E., Hortsch, H., Michler, O., Köhler, T. (eds) *Mobility for Smart Cities and Regional Development - Challenges for Higher Education.* ICL 2021. Lecture Notes in Networks and Systems, vol 389. Springer, Cham. https://doi.org/10.1007/978-3-030-93904-5_56

[7] Medyńska-Gulij, Beata, Jacek Gulij, Paweł Cybulski, Krzysztof Zagata, Jakub Zawadzki, and Tymoteusz Horbiński. 2022. "Map Design and Usability of a Simplified Topographic 2D Map on the Smartphone in Landscape and Portrait Orientations" *ISPRS International Journal of Geo-Information* 11, no. 11: 577. https://doi.org/10.3390/ijgi11110577

[8] Rodríguez-López, Samuel, Matías Ferrán Escobedo Martínez, Luis Junquera, and María García-Pola. 2021. "Two-Dimensional Analysis of Digital Images through Vector Graphic Editors in Dentistry: New Calibration and Analysis Protocol Based on a Scoping Review" *International Journal of Environmental Research and Public Health* 18, no. 9: 4497. https://doi.org/10.3390/ijerph18094497

[9] Martínez-López, J. Israel, Samantha Barrón-González, and Alejandro Martínez López. 2019. "Which Are the Tools Available for Scholars? A Review of Assisting Software for Authors during Peer Reviewing Process" *Publications* 7, no. 3: 59. https://doi.org/10.3390/publications7030059

[10] Diara, Filippo. 2022. "HBIM Open Source: A Review" *ISPRS International Journal of Geo-Information* 11, no. 9: 472. https://doi.org/10.3390/ijgi11090472

[11] Adugna, Y.W., Kurukkal, N.S., and Lemu, H.G. 2024. Anatomical Model Reconstruction (Solid Modeling) Using a Reverse Engineering Approach. In: Pavlou, D., et al. *Advances in Computational Mechanics and Applications*. OES 2023. Structural Integrity, vol 29. Springer, Cham. https://doi.org/10.1007/978-3-031-49791-9_18
[12] Tyupova, Alona, and Joanna Harasym. 2024. "Valorization of Fruit and Vegetables Industry By-Streams for 3D Printing—A Review" *Foods* 13, no. 14: 2186. https://doi.org/10.3390/foods13142186
[13] Oleksy, Małgorzata, Klaudia Dynarowicz, and David Aebisher. 2023. "Rapid Prototyping Technologies: 3D Printing Applied in Medicine" *Pharmaceutics* 15, no. 8: 2169. https://doi.org/10.3390/pharmaceutics15082169
[14] Blender 2.76: Features https://www.blender.org/download/releases/2-76/
[15] Blender 2.76: Download https://download.blender.org/release/Blender2.76/
[16] Blender 2.79b: Features https://www.blender.org/download/releases/2-79/
[17] Blender 2.79: Manual https://docs.blender.org/manual/en/2.79/interface/window_system/introduction.html
[18] Blender 2.79b: Features https://www.blender.org/download/releases/2-79/
[19] Dovramadjiev, Tihomir 2018. 1 Introduction to Blender software. Interface. Zenodo. https://doi.org/10.5281/zenodo.13335546
[20] Dovramadjiev, Tihomir 2018. 2 Introduction to Blender software. 3D View & Editors. Zenodo. https://doi.org/10.5281/zenodo.13335696
[21] Dovramadjiev, Tihomir 2018. 3. Blender software. Editor Type. Zenodo. https://doi.org/10.5281/zenodo.13335733
[22] Dovramadjiev, Tihomir 2018. Blender Tips. How to make display screen bigger or smaller. Zenodo. https://doi.org/10.5281/zenodo.13335711
[23] Dovramadjiev, Tihomir 2018. Blender Tips. How to change language on Russian, German, Spanish, French and other. Zenodo. https://doi.org/10.5281/zenodo.13335765
[24] Dovramadjiev, Tihomir 2018. Blender Tips. How to set Metric or Imperial Units. Zenodo. https://doi.org/10.5281/zenodo.13335783
[25] Dovramadjiev, Tihomir 2018. From Image to Vector to 3D Blender + Inkscape Tutorial. Zenodo. https://doi.org/10.5281/zenodo.13336278
[26] Dovramadjiev, Tihomir 2018. 2D image to 3D scene Blender + Gimp Tutorial. Zenodo. https://doi.org/10.5281/zenodo.13336305
[27] Ragazou, Vasiliki, and Ilias Karasavvidis. 2023. "Effects of Signaling and Practice Types in Video-Based Software Training" *Education Sciences* 13, no. 6: 602. https://doi.org/10.3390/educsci13060602
[28] Feist, Sofia, Luís Jacques de Sousa, Luís Sanhudo, and João Poças Martins. 2024. "Automatic Reconstruction of 3D Models from 2D Drawings: A State-of-the-Art Review" *Eng* 5, no. 2: 784–800. https://doi.org/10.3390/eng5020042
[29] Poku-Agyemang, Kwasi Nyarko, and Alexander Reiterer. 2023. "3D Reconstruction from 2D Plans Exemplified by Bridge Structures" *Remote Sensing* 15, no. 3: 677. https://doi.org/10.3390/rs15030677
[30] Rodriguez-Garcia, Bruno, José Miguel Ramírez-Sanz, Ines Miguel-Alonso, and Andres Bustillo. 2024. "Enhancing Learning of 3D Model Unwrapping through Virtual Reality Serious Game: Design and Usability Validation" *Electronics* 13, no. 10. https://doi.org/10.3390/electronics13101972

7 File Formats (Extensions)

File formats (extensions) are fundamental in 3D digital design, which support standardized means for storing, exchanging, and collaborating on intricate 3D models across a plethora of platforms and software applications (Table 7.1) [1–5]. These formats are crucial in enabling interoperability, which allows designers, engineers, and stakeholders to access and manipulate complex ergonomic designs with high precision and fidelity. They serve as the backbone for storing essential data, including geometric information, textures, materials, and animations that are vital for the development of ergonomic solutions. For instance, widely used formats such as STL (Standard Tessellation Language) and OBJ (Wavefront OBJ) are favored for their simplicity and broad compatibility with various 3D modeling and printing software, with STL being particularly crucial in additive manufacturing, thereby making it indispensable for prototyping ergonomic products. File formats like STEP (Standard for the Exchange of Product Data) and IGES (Initial Graphics Exchange Specification) are exceptional in supporting parametric modeling and geometric data exchange, which is critical for preserving design integrity across different CAD (Computer-Aided Design) systems. These formats enable seamless transfer of complex 3D models between different design tools, facilitating effective collaboration among interdisciplinary teams and precise iteration of ergonomic designs. In ergonomics and human factors, where designs must accommodate anthropometric data and cognitive ergonomics principles, file formats such as DXF (Drawing Exchange Format) and DWG (AutoCAD Drawing Database) are essential. They allow designers to incorporate detailed 2D and 3D representations of ergonomic environments, furniture layouts, and user interfaces, ensuring adherence to ergonomic standards and user-centered design principles. The emerging file formats like glTF (GL Transmission Format) and FBX (Filmbox) are becoming increasingly significant due to their ability to store rich visual and interactive content, including animations and real-time rendering data. These formats are particularly beneficial for virtual reality (VR), augmented reality (AR), and interactive simulations within ergonomics, enhancing user engagement and providing immersive experiences for design evaluations and usability testing. Beyond their technical capabilities, file formats also reflect considerations of accessibility, security, and data integrity. Open-source formats like OBJ and glTF foster transparency and collaboration within the ergonomic design community, encouraging innovation and knowledge sharing. In contrast, proprietary formats such as SolidWorks (SLDPRT, SLDASM) and CATIA (CATPart, CATProduct) offer advanced features and native compatibility with specialized CAD and PLM (Product Lifecycle Management) software, addressing the needs of industries that require strict design control and intellectual property protection [6–12].

DOI: 10.1201/9781003400929-7

TABLE 7.1
Popular Design File Formats

File Format	Extension	Description	Primary Use	Notable Features
STL	`.stl`	Standard Tessellation Language (STL) is a widely used format for 3D printing and additive manufacturing. It represents the surface geometry of a 3D object without any color, texture, or other attributes.	3D Printing, Prototyping	Simplicity, broad compatibility with 3D modeling and printing software.
OBJ	`.obj`	Wavefront OBJ is a neutral format that represents 3D geometry and is compatible with most 3D modeling software. It supports both polygonal geometry and texture mapping, making it versatile for different 3D applications.	3D Modeling, Texturing	Broad compatibility, ability to store geometry, textures, and material properties.
IGES	`.iges, .igs`	Initial Graphics Exchange Specification (IGES) is a file format that allows for the exchange of information among different CAD systems, supporting both 2D and 3D geometry as well as annotations and other elements.	CAD Data Exchange, Legacy CAD Support	Supports complex geometry, annotations, and compatibility with older CAD systems.
DXF	`.dxf`	Drawing Exchange Format (DXF) is a CAD data file format developed by Autodesk. It is used for enabling data interoperability between AutoCAD and other CAD, CAM, and CNC systems, particularly in 2D and 3D design layouts.	2D/3D Design Layouts, CAD Data Exchange	Widely used in architectural and engineering applications, supports 2D and 3D data.
DWG	`.dwg`	AutoCAD Drawing Database (DWG) is a proprietary format used by Autodesk's AutoCAD. It is a native format for CAD designs and supports both 2D and 3D vector graphics, enabling the storage of detailed design information.	CAD Design, Detailed 2D/3D Representations	Comprehensive data storage, including design geometry, annotations, and metadata.
glTF	`.gltf, .glb`	GL Transmission Format (glTF) is a file format for 3D scenes and models that is optimized for use in web applications. It supports rich visual and interactive content, including textures, animations, and real-time rendering data.	Web 3D Applications, VR/AR, Real-time Rendering	Compact size, real-time rendering, and efficient transmission of 3D models over the web.
FBX	`.fbx`	Filmbox (FBX) is a proprietary format used extensively in the film and video game industries for 3D modeling, animation, and content creation. It supports complex 3D models, textures, animations, and lighting data.	Animation, VR/AR, Film, and Video Game Development	Advanced features for animation and real-time interaction, high compatibility with game engines and animation tools.

BLEND	`.blend`	Blender's native file format (.blend) is used to store complete 3D scenes including models, textures, animations, lighting, and more. It is highly versatile and used extensively within the Blender software for all types of 3D projects.	3D Modeling, Animation, Simulation	Comprehensive storage of all project data, including scene settings, mesh data, and more.
SVG	`.svg`	Scalable Vector Graphics (SVG) is a widely used XML-based vector image format for two-dimensional graphics with support for interactivity and animation. It is often used for creating scalable images, icons, and illustrations.	Vector Graphics, Web Design	Scalability without loss of quality, supports interactivity and animation, widely used in responsive design.
SLDPRT/ SLDASM	`.sldprt`, `.sldasm`	SolidWorks Part and Assembly formats are proprietary formats used in SolidWorks CAD software for creating detailed 3D part models and assemblies. These formats are essential for engineering and product development workflows.	Engineering Design, Product Development	Native support for SolidWorks features, integration with PLM systems, and advanced modeling capabilities.
CATPart/ CATProduct	`.catpart`, `.catproduct`	CATIA Part and Product formats are proprietary formats used in CATIA software, widely used in the aerospace, automotive, and manufacturing industries for complex product design and assembly management.	High-end CAD Design, PLM	Advanced features for complex assemblies, integration with industry-specific tools, and support for PLM systems.

The list of file formats and software tools provided in the table highlights some of the most notable and widely used options for 2D and 3D digital designs. While these formats are essential, the list can certainly be extended to include additional formats and tools that may also be useful, depending on specific project requirements and design environments.

REFERENCES

[1] McHenry, Kenton and Bajcsy, Peter (2008) *An Overview of 3D Data Content, File Formats and Viewers*. National Center for Supercomputing Applications University of Illinois at Urbana-Champaign, Urbana, IL. https://www.researchgate.net/publication/228976876_An_overview_of_3D_data_content_file_formats_and_viewers

[2] Alpha3D: 3D file formats https://www.alpha3d.io/3d-file-formats/

[3] All3D: Most common 3D File Formats https://all3dp.com/2/most-common-3d-file-formats-model/

[4] ThePro3DStudio: Popular 3D File Formats Used in Various Industries https://professional3dservices.com/blog/3D-file-formats.html

[5] Adobe: 3D file types https://www.adobe.com/products/substance3d/discover/3d-files-formats.html

[6] Filext: File extensions https://filext.com/

[7] File Info: The File Format Database https://fileinfo.com/

[8] FileFormat: 3D file formats https://docs.fileformat.com/3d/

[9] Dassault Systèmes: SolidWorks DXF/DWG Files (*.dxf, *.dwg Files) https://help.solidworks.com/2024/english/SolidWorks/sldworks/c_DXF-DWG_Files_dxf_dwg-Files.htm

[10] Dassault Systèmes: SolidWorks File Types https://help.solidworks.com/2024/english/SolidWorks/sldworks/c_File_Types.htm

[11] Cad Infield: SOLIDWORKS File Extension https://cadinfield.com/solidworks-file-extension/

[12] Capvidia Blog (2023) Top 8 Neutral 3D CAD File Formats https://www.capvidia.com/blog/top-neutral-3d-cad-file-formats

8 Human 3D Digital Design

For 3D digital human design and E/HF, the **Open Anatomy Project** is an invaluable resource that greatly enhances the precision and effectiveness of digital design practices [1]. By providing free, interactive digital anatomy atlases—such as the SPL/NAC Brain Atlas and the SPL Knee Atlas—the project offers detailed, three-dimensional visualizations of the human body, which are essential for both healthcare-related and conventional design applications. These tools enable designers and engineers to explore intricate anatomical structures and their spatial relationships, crucial for creating ergonomic products, virtual characters, and human-centered interfaces in various industries. The project's commitment to open access fosters a collaborative environment, allowing professionals across disciplines to leverage high-quality anatomical data for optimizing user comfort, functionality, and overall design efficacy. By integrating these detailed digital resources, the Open Anatomy Project supports the development of designs that are both anatomically accurate and ergonomically relevant, advancing the field of 3D digital human design and enhancing user experience in diverse applications.

Another free open resource is **OpenStax Anatomy and Physiology**, which stands out as a vital tool for 3D digital human design and ergonomics [2]. Developed by OpenStax, this meticulously crafted educational textbook offers a thorough exploration of human anatomy and physiology, covering a broad spectrum of topics from foundational concepts to complex physiological systems. Its six comprehensive units delve into critical areas such as levels of organization, support and movement, regulation and control, fluids and transport, energy maintenance, and human development. The textbook's interactive features, including online exercises, simulations, and videos, enhance learning by providing dynamic visualizations and engaging content. This open-access resource not only supports the creation of anatomically accurate 3D models and E/HF designs but also allows for customization to meet specific educational and design needs. By integrating OpenStax Anatomy and Physiology into design workflows, professionals and educators can achieve a deeper understanding of human anatomy and its applications, leading to more effective and user-centered designs.

For more understanding and well-visible interaction, the developed free interactive software **Anatronica** offers an exceptional resource for 3D digital human design and anthropometry [3]. This advanced platform provides a highly immersive 3D experience, allowing users to manipulate and explore detailed anatomical models with precision and ease. Anatronica's robust features include a comprehensive Bone Learn Database, which delivers real-time information about specific bones, and interactive quizzes that reinforce learning through immediate feedback and detailed statistics (Figure 8.1). These elements enhance the comprehension of complex

DOI: 10.1201/9781003400929-8

FIGURE 8.1 Anatronica Interactive 3D Human Anatomy software interface (left) Theory; (right) Quiz result example. DOI: 10.6722/TES.202404_8(1).0005 [4].

anatomical relationships and spatial dynamics, which are essential in fields such as ergonomics, implantology, and bioengineering. By facilitating hands-on learning and promoting interdisciplinary collaboration, Anatronica not only supports the development of accurate and ergonomic designs but also advances innovation in healthcare solutions. Its free nature ensures that high-quality educational resources are widely available, empowering users to deepen their anatomical knowledge and improve their design practices in both medical and conventional contexts.

By utilizing Anatronica's real-time access to bone-specific details, users can explore complex anatomical structures with greater clarity. The software's interactive quizzes provide immediate feedback, allowing users to gauge their comprehension and refine their understanding of bone anatomy. The development of video Bone Anatomical Presentations, covering topics such as skull, teeth, hands, body, and legs, further enriches the learning experience by offering dynamic visualizations. These tools are instrumental in advancing 3D digital human design, enabling more precise and effective ergonomic solutions. For example, through the integration of Anatronica into the ErgoDesign project, software not only supports the educational needs of students and stakeholders but also drives innovation in ergonomic design by providing a comprehensive and interactive approach to anatomical education. The file: Additional Resources (Table 8.1).

In practice, human 3D digital design involves a diverse array of methodologies and software tools tailored to specific ergonomic applications. This iterative process allows designers to refine ergonomic attributes such as posture, reach, and visibility, while also addressing cognitive factors like decision-making, workload, and user interface design, ensuring solutions that adhere to human-centered design principles. Human 3D digital design fosters interdisciplinary collaboration by facilitating communication among stakeholders from various sectors, including healthcare, automotive, aerospace, and consumer products. Feedback from ergonomists, psychologists, engineers, and end-users integrates into digital simulations, driving innovation and user-centricity in product development and workplace design. The use of simulation-based training and virtual prototyping enables cost-effective evaluation of ergonomic interventions, addressing potential issues before physical implementation. Looking forward, advancements in artificial intelligence, machine learning, and digital twin technologies are set to enhance the accuracy and predictive capabilities of digital human models, enabling adaptive ergonomic solutions tailored to individual preferences and environmental conditions. The rise of cloud-based platforms and open-source frameworks democratizes access to these tools, empowering designers worldwide to tackle complex ergonomic challenges in an increasingly interconnected world.

8.1 HUMAN 3D DIGITAL DESIGN WITH OSS BLENDER (+ MB-LAB)

Human 3D digital design using open-source software (OSS) Blender, combined with MB-Lab, represents a technological advancement in the field of ergonomics and human factors. Blender, renowned for its robust capabilities in 3D modeling, animation, and rendering, empowers designers and researchers to create highly detailed and realistic human models tailored specifically for ergonomic studies. MB-Lab, an open-source add-on for Blender, enhances this capability by offering a streamlined workflow for generating customizable human characters with precise anatomical features and proportions. This integration facilitates the creation of digital avatars that accurately simulate human interactions with various environments, tools, and technologies, crucial for evaluating ergonomic designs and interventions. Blender's open-source nature fosters collaboration and innovation within the ergonomics community, enabling researchers worldwide to freely access and contribute to the development of sophisticated digital human models. By leveraging Blender and MB-Lab, practitioners in ergonomics and human factors can conduct comprehensive analyses, optimize workplace designs, mitigate ergonomic risks, and ultimately enhance human well-being and performance in diverse settings from healthcare to industrial environments. This intersection of advanced digital design tools with ergonomic principles underscores their transformative impact on shaping future-oriented solutions that promote safety, efficiency, and user-centric design across global industries and societies.

- **MB-Lab**, formerly known as ManuelBastioniLAB, is a renowned free Blender add-on celebrated for its ability to generate diverse 3D human characters effortlessly. Initially developed for Blender 2.7x, it has evolved through subsequent versions to now support Blender 4. The add-on offers a wide range of character phenotypes, including Asian, Caucasian, Afro, and others catering to various artistic needs. Although development was taken over by the community after the original developer's cessation, MB-Lab continues to expand its capabilities. Recent updates, such as the introduction of the MB-Dev Character Development Framework and enhancements to the Hair Engine and Material Engine, highlight its ongoing evolution and commitment to robust character creation within Blender's system [5].

MB-Lab integrates seamlessly into Blender's interface, offering an intuitive GUI optimized for user-friendly operation. Key features include an innovative control system with age, body mass, and tone sliders that define over 90% of character attributes, supported by advanced tools for detailed customization of body and facial features, skin shaders, and animation. Written in Python, MB-Lab leverages cutting-edge technologies such as nonlinear interpolation and fuzzy mathematics to refine character parameters like age progression and body proportions. The plugin supports diverse platforms, including Windows, macOS, and Linux; adheres to the AGPL and GPL licenses; and provides extensive anatomical detail with optimized mesh topologies for professional animation and sculpting. With its comprehensive library of phenotypes representing Caucasian, Asian, Afro, and other human variations, MB-Lab remains a very useful tool for artists and developers seeking to create lifelike 3D human models and beyond in Blender environments. MB-Lab, an add-on for Blender, supports Blender 2.79 (version MB-LAB 1.6.5, free for download: https://github.com/animate1978/MB-Lab/archive/1.6.5b.zip, Table 8.2) and the latest version for Blender 4.0 (MB-LAB 1.8.0, free for download: https://github.com/animate1978/MB-Lab/archive/refs/heads/master.zip). The source codes are placed for Blender

TABLE 8.2
Blender Version 2.79b and ManuelBastioniLAB Version 1.6.1

Steps	Install MB-Lab ManuelBastioniLAB version 1.6.1 to Blender 2.79b	Display Configuration
1	Start Blender software > File > User Preferences > Add-ons > Install Add-on from file > MB-Lab file	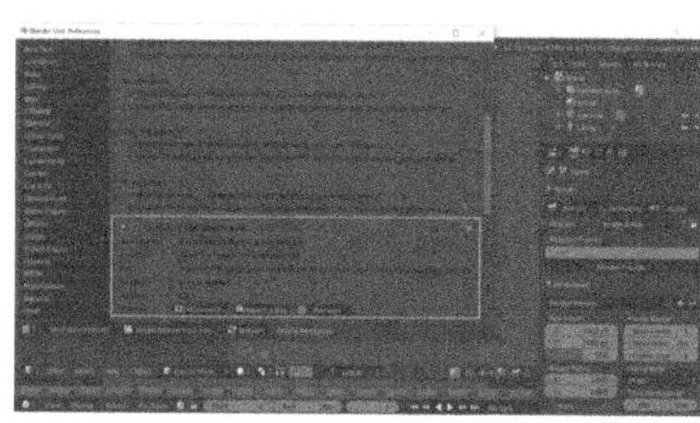
2	Init character	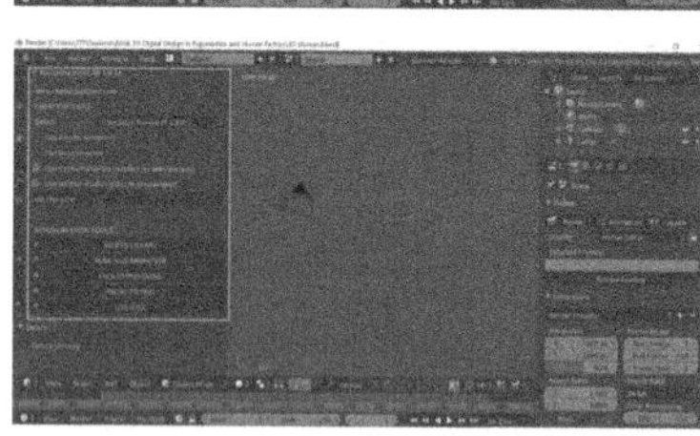
3	Default parameters: Caucasian female (F_CA01). Age: 33 y; Mass: 50%; Tone 50%. Character age: 0.000; Character Mass: 0.000; Character tone: 0.000.	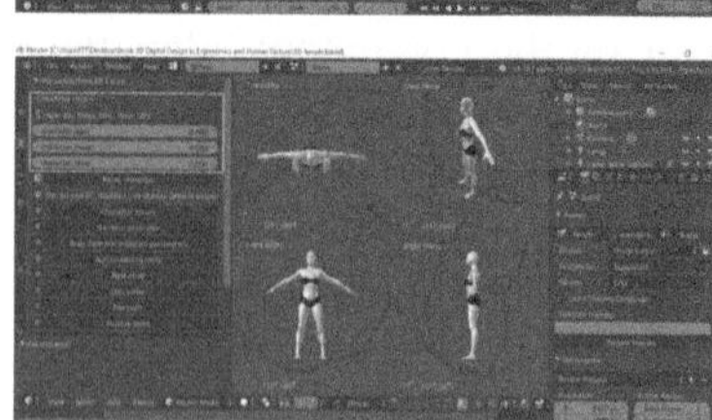

2.79: https://github.com/BlenderNewbie2020/MB-Lab-2.79 and for Blender 4.0: https://github.com/animate1978/MB-Lab. MB-Lab is a robust open-source add-on for Blender, licensed under the GNU Affero General Public License Version 3 (AGPL-3.0) for data files and the GNU General Public License 3 (GPL-3.0) for Python files, ensuring that it remains free and accessible to all. This dual licensing allows users to freely copy, modify, and distribute the software, with guaranteed access to its source code and data. Developed by a global community of artists and programmers, MB-Lab is designed for professional 3D character modeling, offering advanced features like optimized topology for subdivision surfaces and sculpting, integrated elements such as teeth and eyes, and a robust, animation-ready rigging system. It includes an extensive library of anthropological phenotypes, advanced shaders for Cycles and EEVEE, and a proxy fitting system for adding clothing and hair. With its powerful morphing capabilities, anime-style character support, and comprehensive facial expression library, MB-Lab is an essential tool for creating detailed, realistic, and expressive 3D characters. [6, 7]. **MB-Lab's base characters** are meticulously designed to meet professional standards for diverse 3D applications. The mesh topology is optimized for subdivision surfaces, ensuring efficient polygon use and natural deformation during animation, which is crucial for creating detailed facial expressions and realistic body movements. Each character model includes fully integrated teeth and eye models, ensuring compatibility across various platforms without relying on additional systems. Derived from extensive anatomical and anthropological research, these models represent idealized average traits for each phenotype, ensuring accuracy and realism. The character creation process in MB-Lab is divided into two key phases: creation and post-creation. During creation, users define essential character attributes such as age, muscle tone, and facial features using intuitive parametric tools. In the post-creation phase, the character becomes a standard model ready for advanced tasks like animation, expression editing, and the integration of assets such as clothing and hair. MB-Lab's tools enable detailed customization, making it an essential resource for producing high-quality, animation-ready characters [8–11]. MB-Lab offers several methods **for saving or exporting characters**, each suited to different needs. Finalizing the character in Blender converts it into a standard .blend file, fully rigged and ready for further editing, but this process makes MB-Lab-specific features irreversible. Alternatively, saving the .blend file without finalizing preserves these features, though re-initialization may fail if modifications like shape keys are added. A more flexible option is exporting the character as JSON, which retains all MB-Lab parameters and features, ensuring easy upgrades, conversions, and restoration across different environments and MB-Lab versions [12].

Table 8.3 shows the process of 3D digital design of Caucasian female characters. Similar processes are shown in Table 8.4 Asian female character, Table 8.5 Afro female character.

Figure 8.2 shows the final render image (Cycles) of a ready 3D model of a Caucasian female character. The technical parameters include Left rendered image: Samples: Render: 128 Preview: 32; Right rendered image: Samples: Render: 1000 Preview: 32.

Figure 8.3 shows the creation of a 3D digital model of a Caucasian male.

TABLE 8.3
3D Digital Design of Caucasian Female Character

Character: Caucasian Female (F_CA01) Parameters:	3D Digital Human Visualization (Blender software + MB-LAB add-on)
Age: 18 y Mass: 50% Tone: 50%. Character age: -1.000 Character Mass: 0.000 Character tone: 0.000	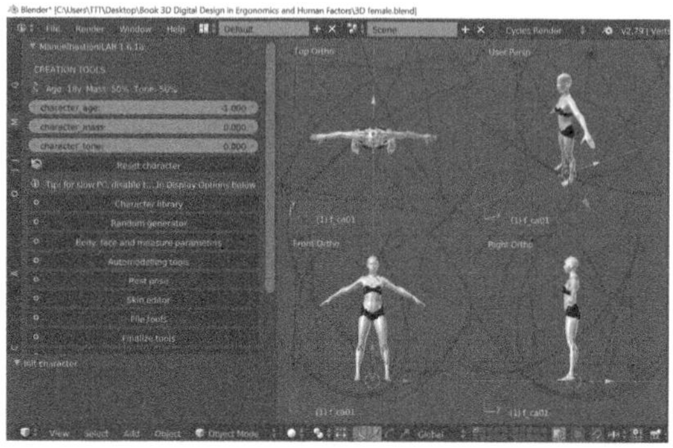
Age: 21 y Mass: 50% Tone: 50%. Character age: -0.500 Character Mass: 0.000 Character tone: 0.000	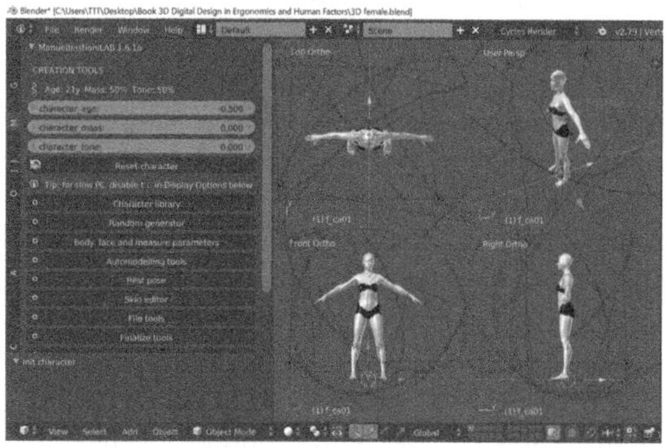
Age: 52 y Mass: 50% Tone: 50%. Character age: 0.500 Character Mass: 0.000 Character tone: 0.000	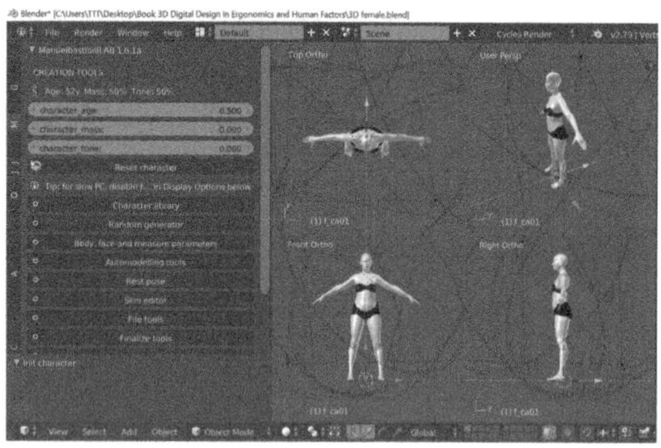

(Continued)

TABLE 8.3
(Continued)

Character: Caucasian Female (F_CA01) Parameters:	3D Digital Human Visualization (Blender software + MB-LAB add-on)
Age: 80 y Mass: 50% Tone: 50%. Character age: 1.000 Character Mass: 0.000 Character tone: 0.000	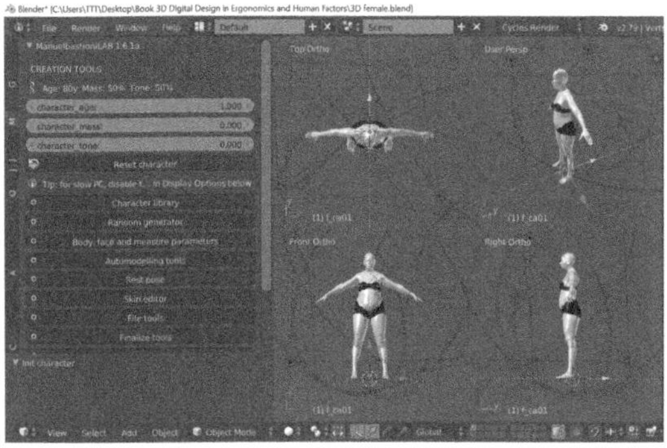

TABLE 8.4
3D Digital Design of Asian Female Character

Character: Asian Female (F_AS01) Parameters:	3D Digital Human Visualization (Blender Software + MB-LAB Add-on)
Age: 18 y Mass: 50% Tone: 50%. Character age: -1.000 Character Mass: 0.000 Character tone: 0.000	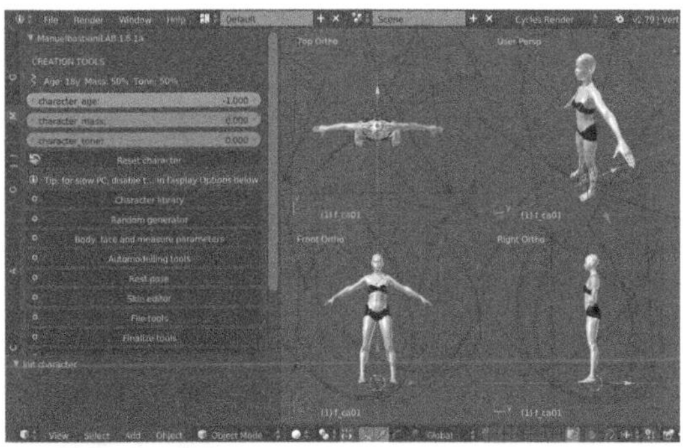
Age: 21 y Mass: 50% Tone: 50%. Character age: -0.500 Character Mass: 0.000 Character tone: 0.000	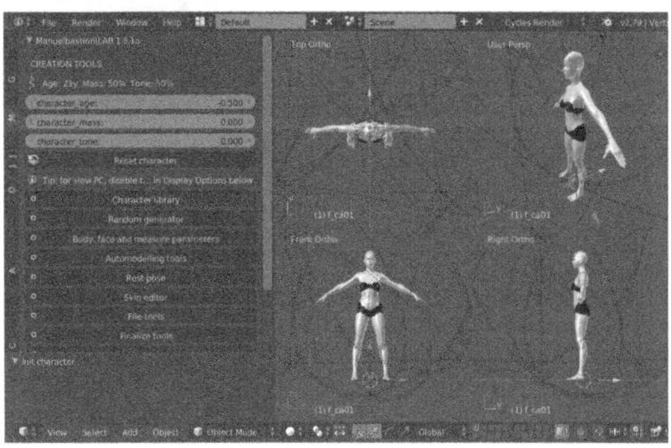

(Continued)

TABLE 8.4 (Continued)

Character: Asian Female (F_AS01) Parameters:	3D Digital Human Visualization (Blender Software + MB-LAB Add-on)
Age: 33 y Mass: 50% Tone: 50%. Character age: 0.000 Character Mass: 0.000 Character tone: 0.000	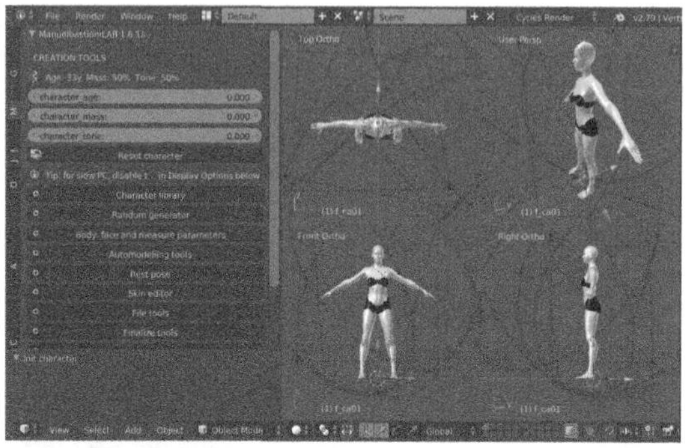
Age: 52 y Mass: 50% Tone: 50%. Character age: 0.500 Character Mass: 0.000 Character tone: 0.000	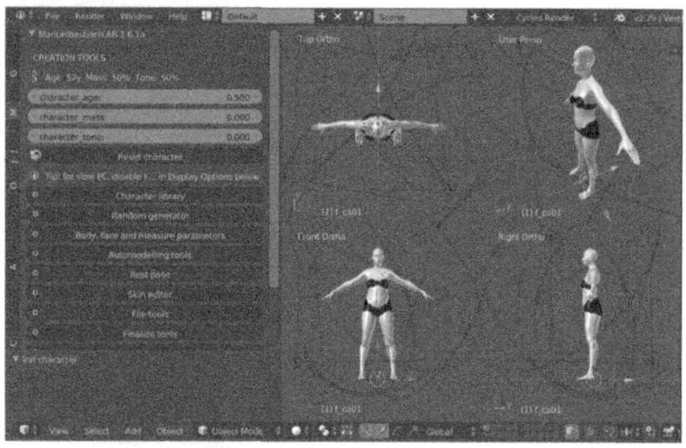
Age: 80 y Mass: 50% Tone: 50%. Character age: 1.000 Character Mass: 0.000 Character tone: 0.000	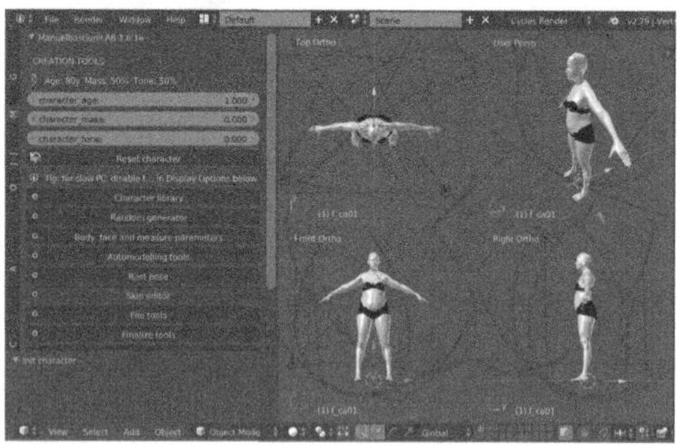

TABLE 8.5
3D Digital Design of Afro Female Character

Character: Afro Female (F_AF01) Parameters:	3D Digital Human Visualization (Blender Software + MB-LAB add-on)
Age: 18 y Mass: 50% Tone: 50%. Character age: -1.000 Character Mass: 0.000 Character tone: 0.000	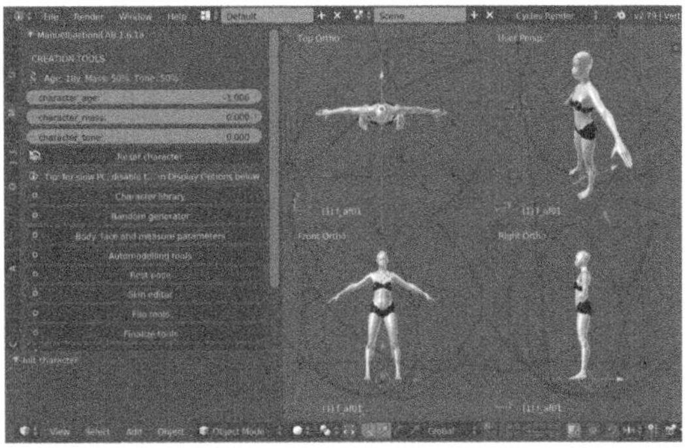
Age: 21 y Mass: 50% Tone: 50%. Character age: -0.500 Character Mass: 0.000 Character tone: 0.000	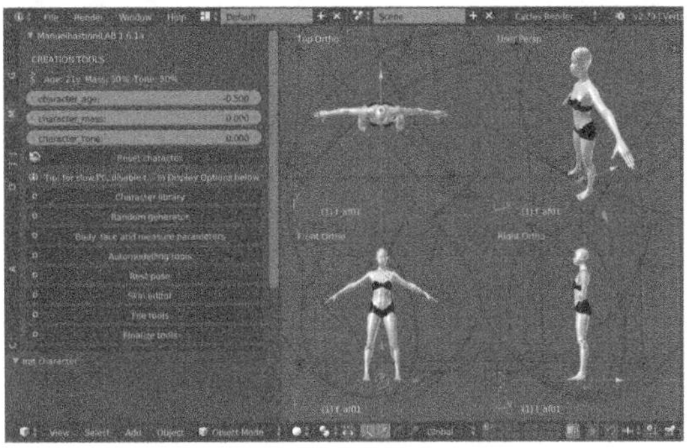
Age: 33 y Mass: 50% Tone: 50%. Character age: 0.000 Character Mass: 0.000 Character tone: 0.000	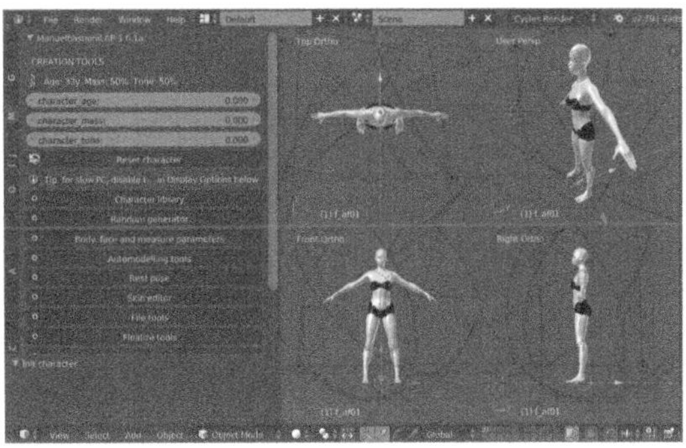

(Continued)

TABLE 8.5
(Continued)

Character: Afro Female (F_AF01) Parameters:	3D Digital Human Visualization (Blender Software + MB-LAB add-on)
Age: 52 y Mass: 50% Tone: 50%. Character age: 0.500 Character Mass: 0.000 Character tone: 0.000	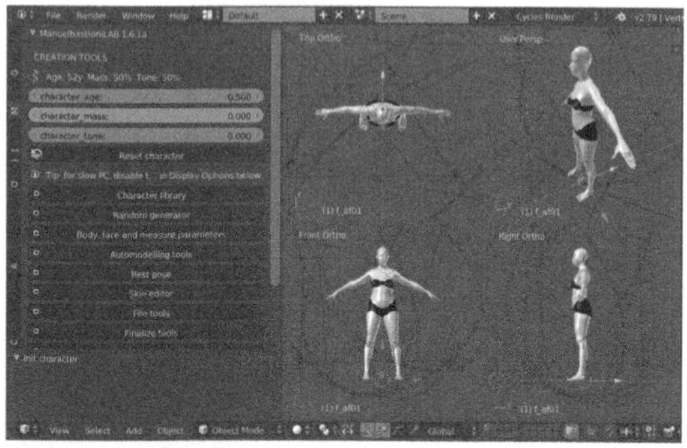
Age: 80 y Mass: 50% Tone: 50%. Character age: 1.000 Character Mass: 0.000 Character tone: 0.000	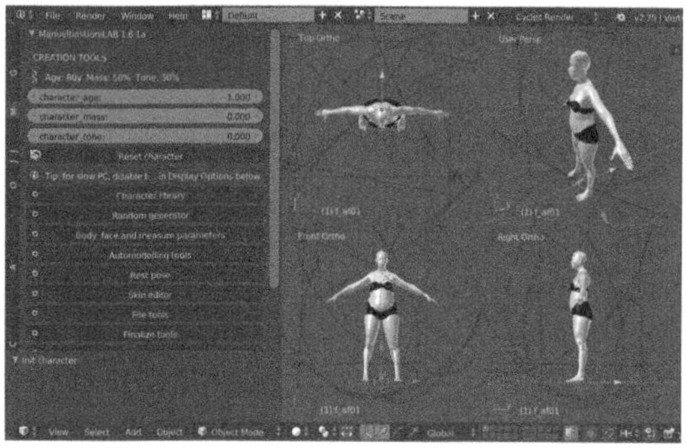

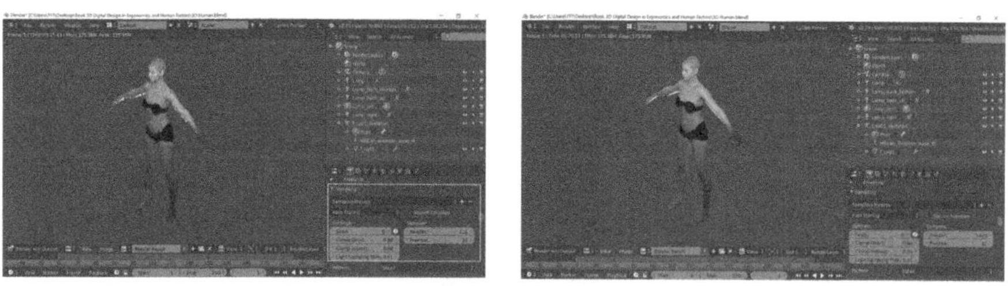

FIGURE 8.2 Blender Cycles rendered character of Caucasian female (left—low quality) default system parameters; (right—high quality) Samples: Render: 1000

Similar to female characters, Table 8.6 shows the process of 3D digital design of a Caucasian male character. Similar processes are shown in Table 8.7, Asian male character, and Table 8.8, Afro male character.

Also, similar to the female character, Figure 8.4 shows the final render image (Cycles) of the ready 3D model of a Caucasian male character. The technical parameters include Left rendered image: Samples: Render: 128 Preview: 32; Right rendered image: Samples: Render: 1000 Preview: 32.

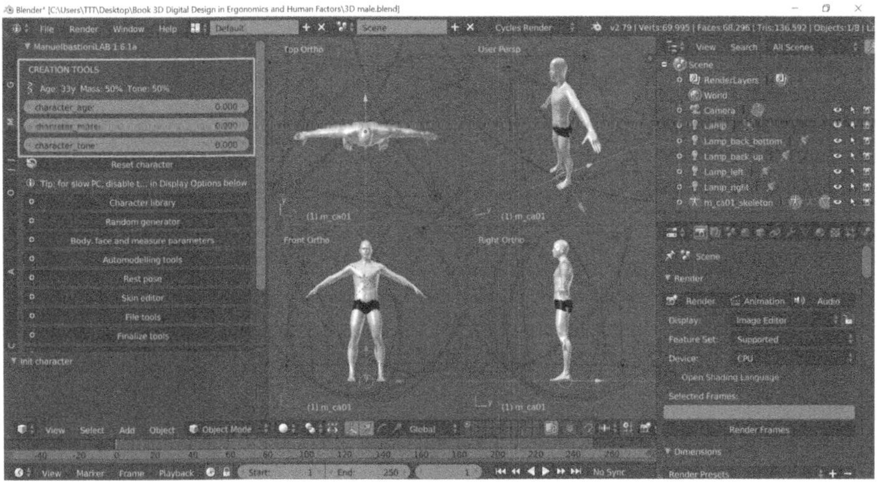

FIGURE 8.3 MB-Lab: Default parameters: Caucasian male (M_CA01). Age: 33 y; Mass: 50%; Tone: 50%. Character age: 0.000; Character Mass: 0.000; Character tone: 0.000.

TABLE 8.6
3D Digital Design of Caucasian Male Character

Character: Caucasian Male (M_CA01) Parameters:	3D Digital Human Visualization (Blender Software + MB-LAB Add-on)
Age: 18 y Mass: 50% Tone: 50%. Character age: -1.000 Character Mass: 0.000 Character tone: 0.000	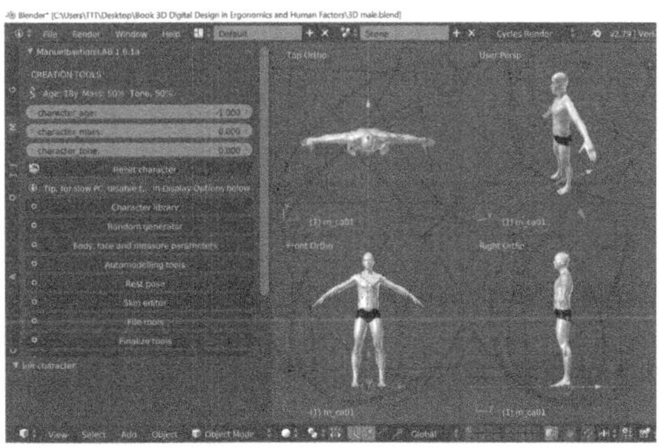
Age: 21 y Mass: 50% Tone: 50%. Character age: -0.500 Character Mass: 0.000 Character tone: 0.000	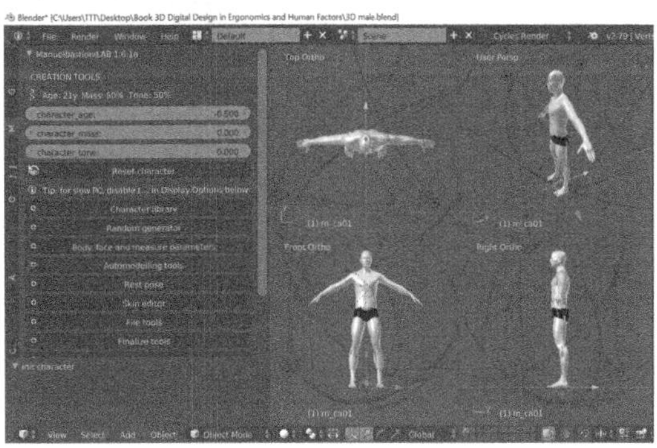

(Continued)

TABLE 8.6
(Continued)

Character: Caucasian Male (M_CA01) Parameters:	3D Digital Human Visualization (Blender Software + MB-LAB Add-on)
Age: 52 y Mass: 50% Tone: 50%. Character age: 0.500 Character Mass: 0.000 Character tone: 0.000	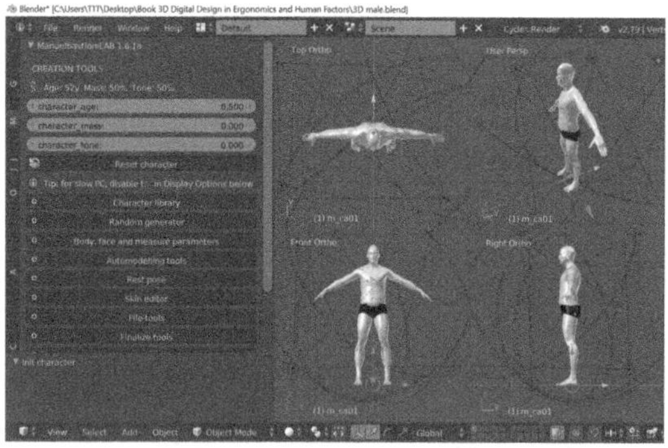
Age: 80 y Mass: 50% Tone: 50%. Character age: 1.000 Character Mass: 0.000 Character tone: 0.000	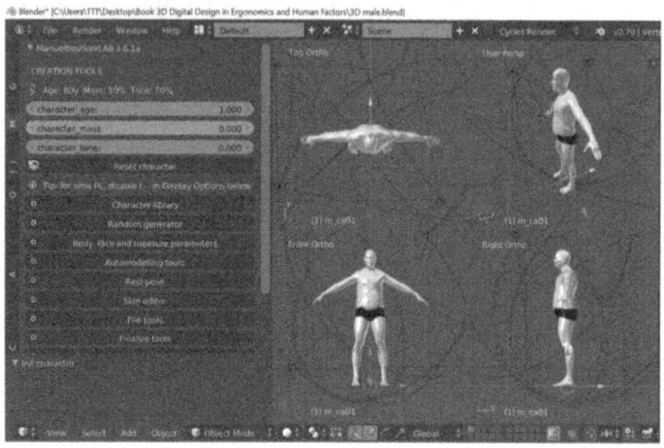

TABLE 8.7
3D Digital Design of Asian Male Character

Character: Asian Male (M_AS01) Parameters:	3D Digital Human Visualization (Blender Software + MB-LAB Add-on)
Age: 18 y Mass: 50% Tone: 50%. Character age: -1.000 Character Mass: 0.000 Character tone: 0.000	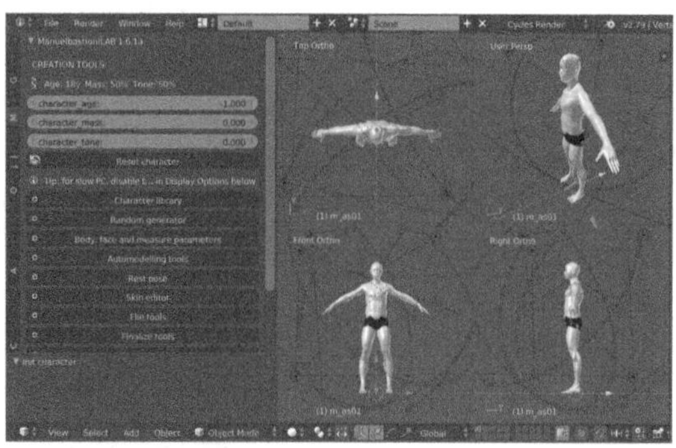

(Continued)

TABLE 8.7
(Continued)

Character: Asian Male (M_AS01) Parameters:	3D Digital Human Visualization (Blender Software + MB-LAB Add-on)
Age: 21 y Mass: 50% Tone: 50%. Character age: -0.500 Character Mass: 0.000 Character tone: 0.000	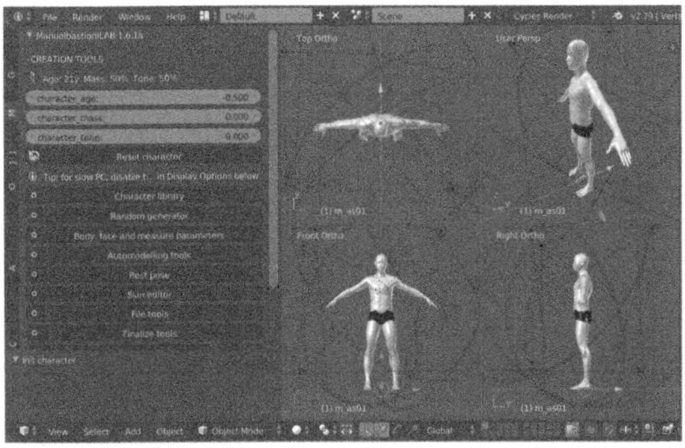
Age: 33 y Mass: 50% Tone: 50%. Character age: 0.000 Character Mass: 0.000 Character tone: 0.000	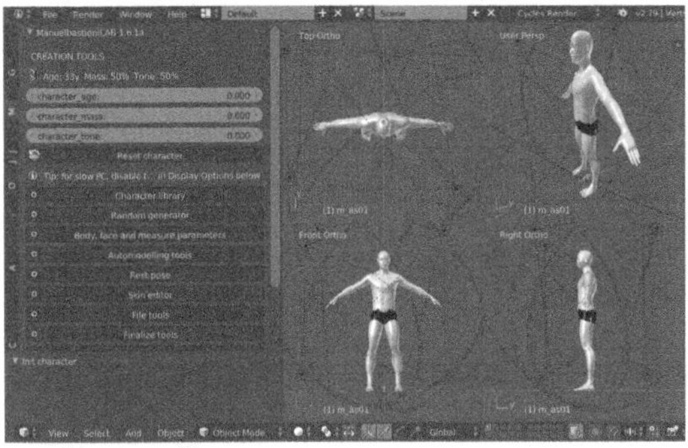
Age: 52 y Mass: 50% Tone: 50%. Character age: 0.500 Character Mass: 0.000 Character tone: 0.000	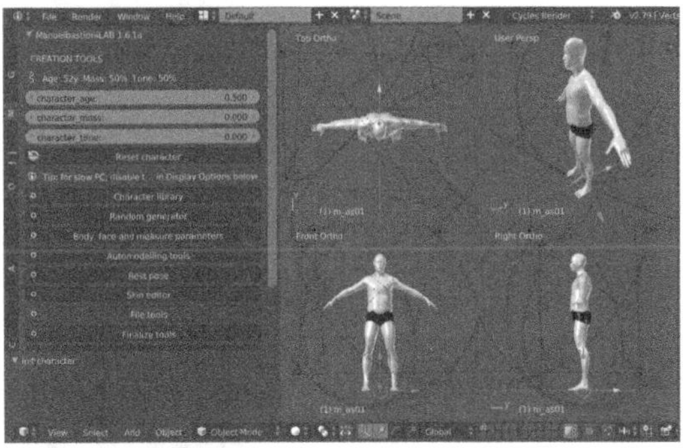
Age: 80 y Mass: 50% Tone: 50%. Character age: 1.000 Character Mass: 0.000 Character tone: 0.000	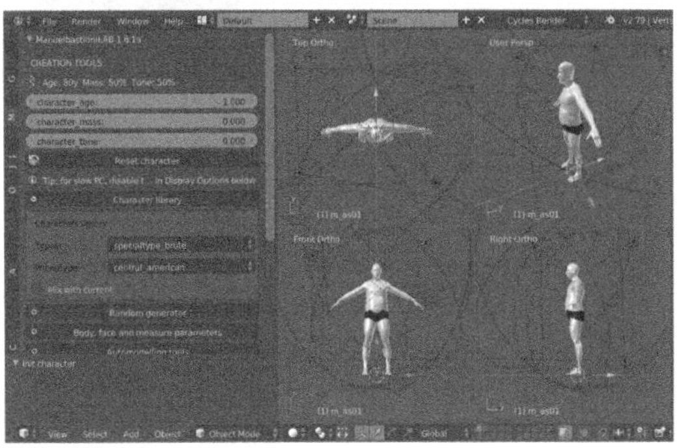

TABLE 8.8
3D Digital Design of Afro Male Character

Character: Afro Male (M_AF01) Parameters:	3D Digital Human Visualization (Blender Software + MB-LAB Add-on)
Age: 18 y Mass: 50% Tone: 50%. Character age: -1.000 Character Mass: 0.000 Character tone: 0.000	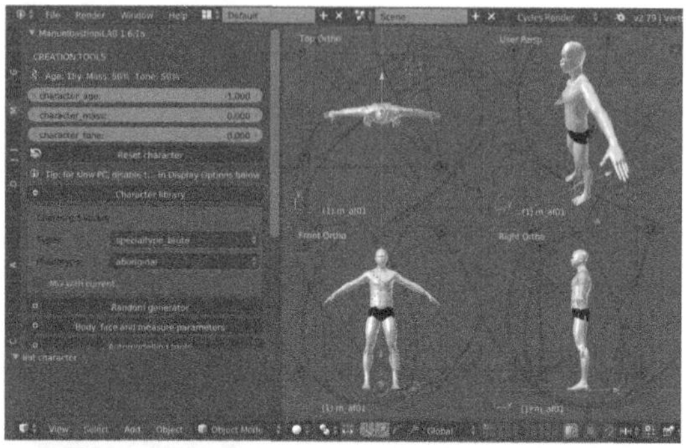
Age: 21 y Mass: 50% Tone: 50%. Character age: -0.500 Character Mass: 0.000 Character tone: 0.000	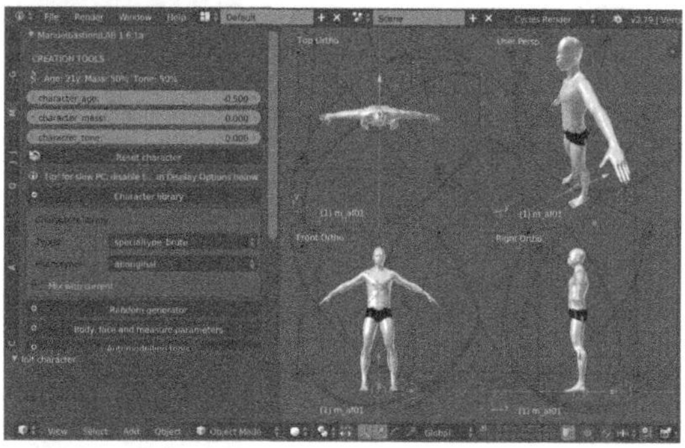
Age: 33 y Mass: 50% Tone: 50%. Character age: 0.000 Character Mass: 0.000 Character tone: 0.000	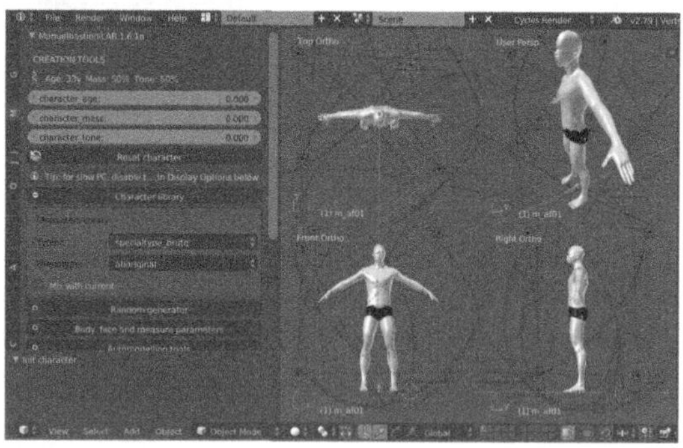

(*Continued*)

TABLE 8.8
(Continued)

Character: Afro Male (M_AF01) Parameters:	3D Digital Human Visualization (Blender Software + MB-LAB Add-on)
Age: 52 y Mass: 50% Tone: 50%. Character age: 0.500 Character Mass: 0.000 Character tone: 0.000	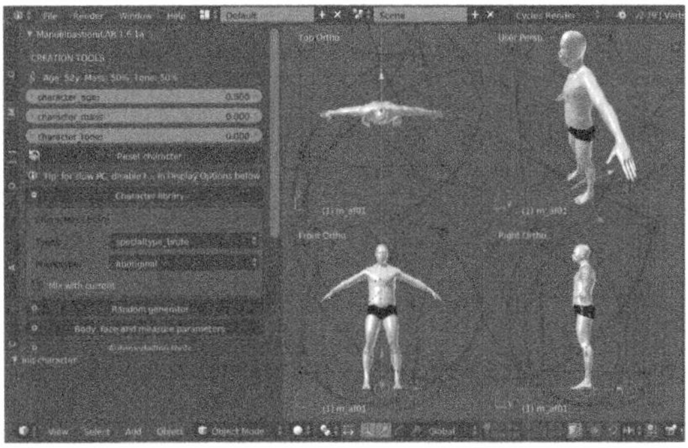
Age: 80 y Mass: 50% Tone: 50%. Character age: 1.000 Character Mass: 0.000 Character tone: 0.000	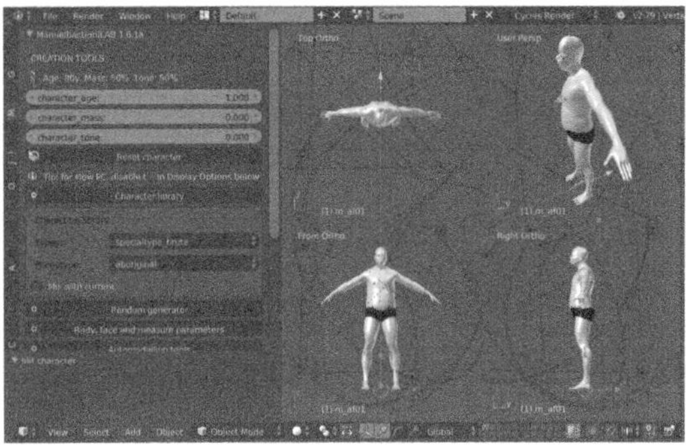

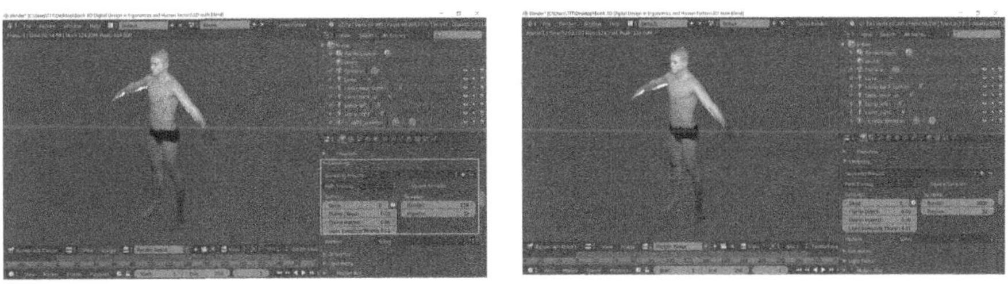

FIGURE 8.4 Blender Cycles rendered character of Caucasian male (left—low quality) default system parameters; (right—high quality) Samples: Render: 1000.

MB-Lab's Creation Tools offer a robust system for customizing human models through various parameters and presets. After placing a basic human model in the Blender scene, users can manipulate it using six methods: meta parameters (age, mass, and body tone), character library presets, quasi-random generators, auto-modeling tools, low-level parameter adjustments, and measurement tools. The preset library allows for quick character definition by loading types based on anthropometric data and artistic observations, which define features like bone structure, fat percentage, mass, and muscle. These types, labeled intuitively as "ideal fashion," "athletic," etc., make the system user-friendly. Ethnic phenotypes from various world regions—Afro, Asian, Caucasian, and Latino—are included, allowing for diverse character creation reflective of global anthropometric variations. This comprehensive set of tools enables detailed customization and realistic character modeling with ease (Table 8.9, 3D female character, and Table 8.10, 3D male character).

MB-Lab's library types for male and female models are distinct, with each gender's types reflecting specific anthropometric features and body shapes tailored to realistic and diverse human forms.

TABLE 8.9
3D Digital Design of Type and Phenotype of Female Character

Character Library/ Parametric Design	3D Female
Type: normobody Phenotype: central_european	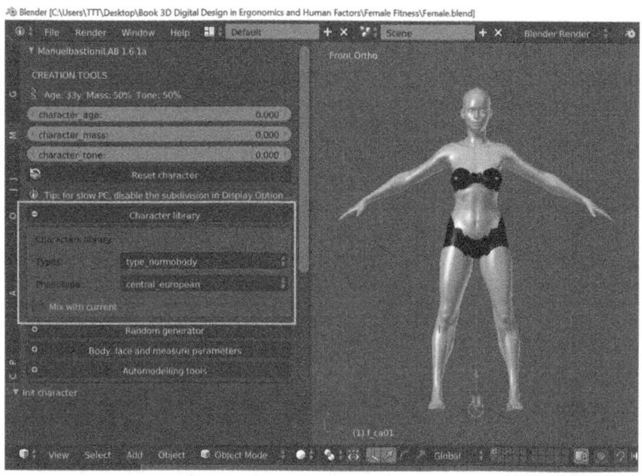
Types: by choice Phenotype:	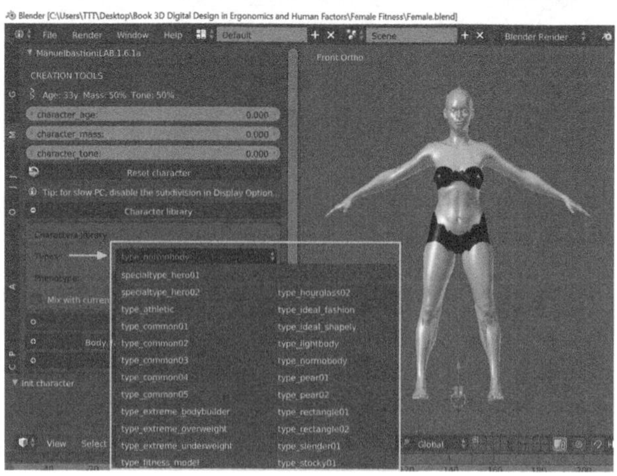

(Continued)

TABLE 8.9 (Continued)

Character Library/ Parametric Design	3D Female
Types: Phenotype: by choice	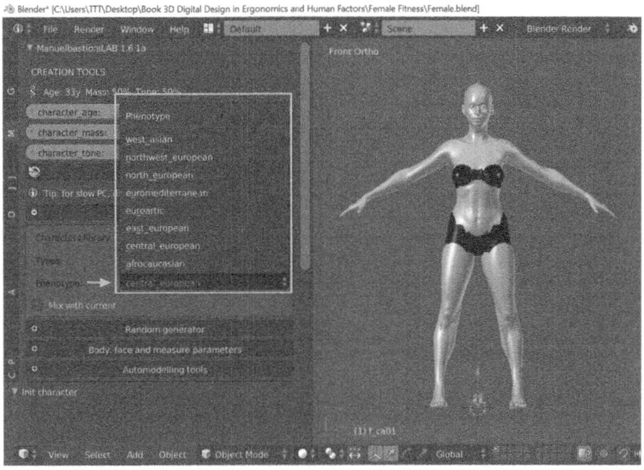
Type: type_athletic Phenotype: central_european	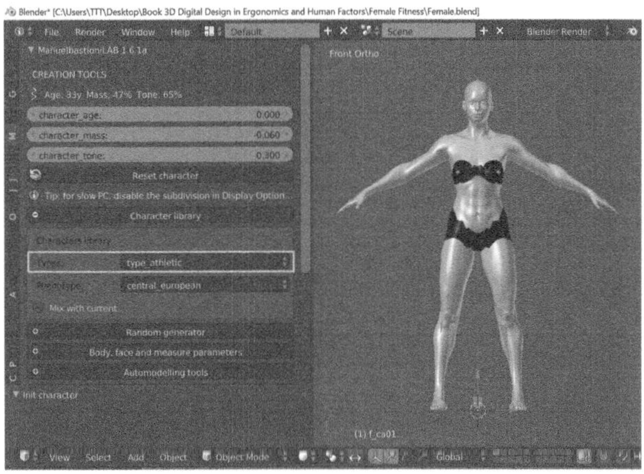
Type: type_fitness_model Phenotype: central_european	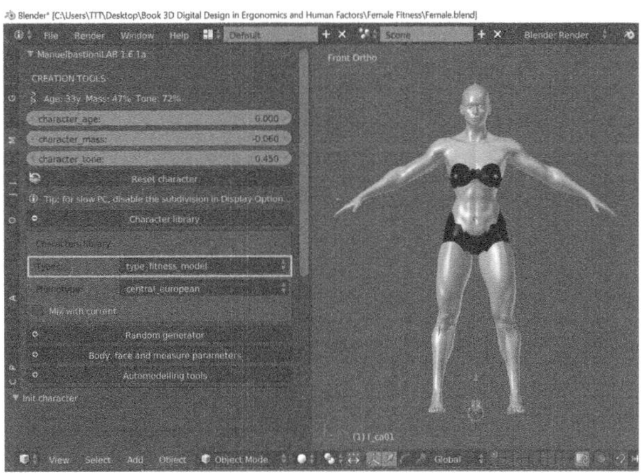

(Continued)

TABLE 8.9
(Continued)

Character Library/ Parametric Design	3D Female
Type: type_ideal_fashion Phenotype: central_european	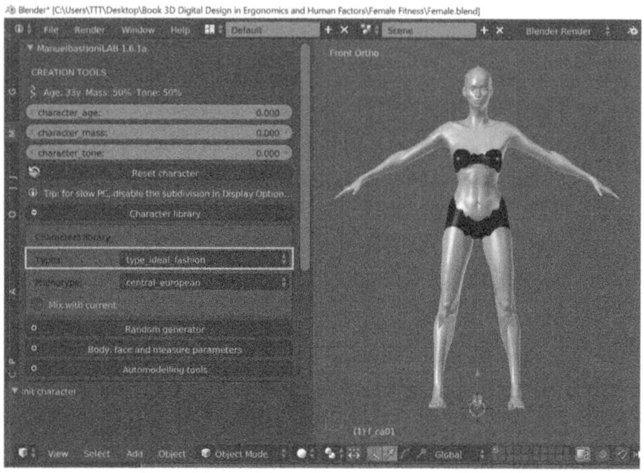
Type: type_ideal_shapely Phenotype: central_european	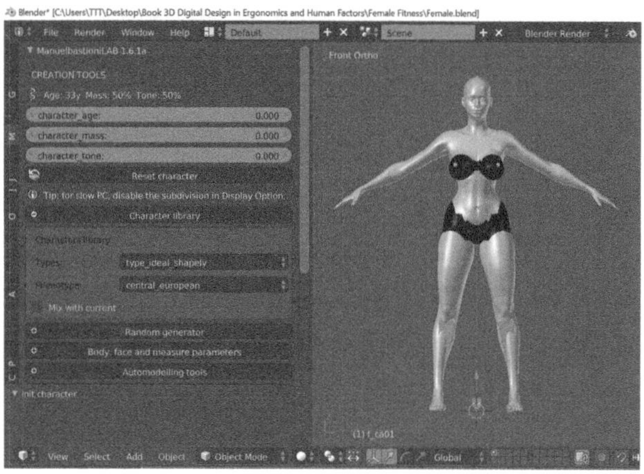

TABLE 8.10
3D Digital Design of Type and Phenotype of Male Character

Character Library/ Parametric Design	3D Male
Type: normobody Phenotype: central_european	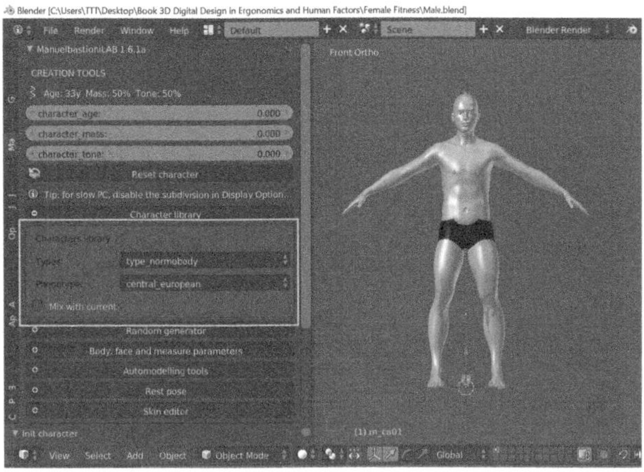
Types: by choice Phenotype:	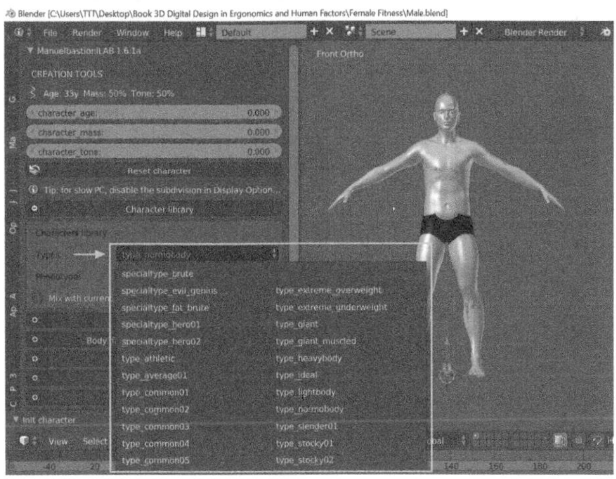

(Continued)

TABLE 8.10 (Continued)

Character Library/ Parametric Design	3D Male
Types: Phenotype: by choice	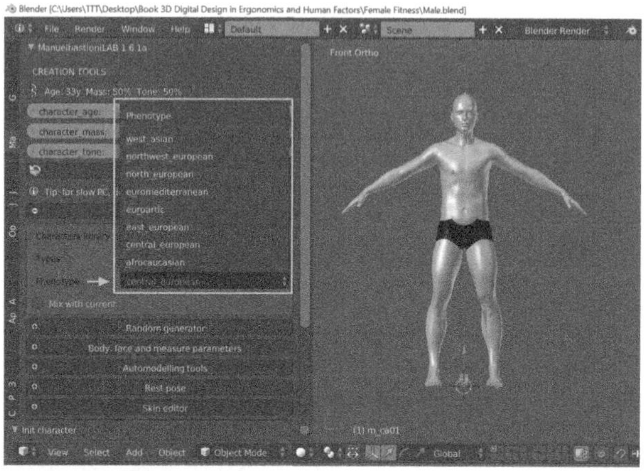
Type: type_athletic Phenotype: central_european	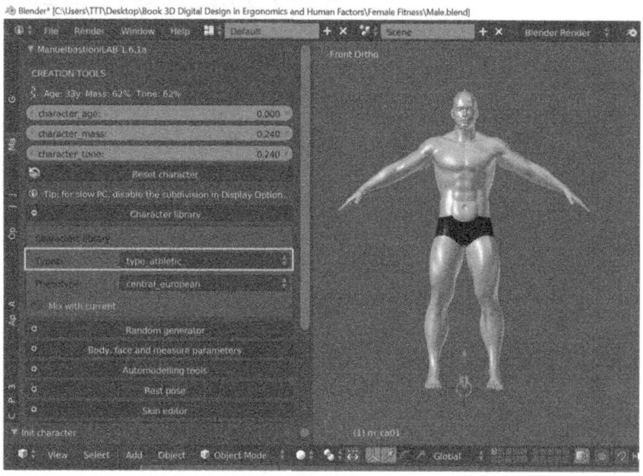
Type: type_ideal Phenotype: central_european	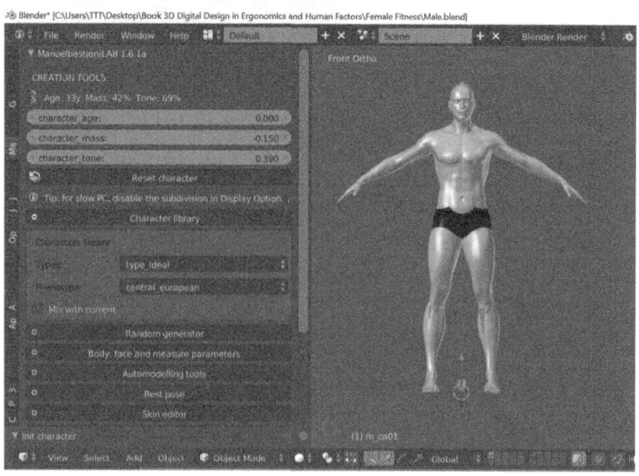

8.2 HUMAN 3D DIGITAL DESIGN WITH OSS MAKEHUMAN

MakeHuman (an OSS) is of high importance for the development of ergonomics and advanced design in any environment where human factors are integral, providing high-quality human digital models essential for simulations and understanding human interactions in real-world scenarios. This open-source software significantly impacts 3D digital design by offering an extensive suite of tools for creating detailed and customizable human models, crucial for accurately representing anthropometric dimensions. This capability is essential for optimizing designs across various applications, including workplace ergonomics and product development. MakeHuman's intuitive interface and robust parametric controls allow for precise adjustments of body proportions, facial expressions, and poses, enabling effective virtual simulations and ergonomic evaluations. The software's integration with Blender facilitates smooth data transfer and further refinement of 3D models, enhancing workflow efficiency. Notable MakeHuman versions like 1.1.1 [13] and 1.3.0 [14] bring significant improvements, including better stability, compatibility with modern operating systems, and new features. This ongoing development and support highlight MakeHuman's role in producing realistic human models and advancing ergonomic design, fostering innovation and collaboration within the design community. One of the most important features of MakeHuman software is its strong integration with Blender software, facilitating seamless transfer and further refinement of 3D human digital models. Despite the seemingly incremental updates, it is crucial for stakeholders and designers to appreciate and respect the continuous efforts of developers who work tirelessly to enhance and maintain the software, ensuring it remains a valuable tool for the 3D digital design community [15]. The development process of the completed 3D digital model of female and male is shown in Tables 8.11 (female) and Table 8.12 (male).

MakeHuman and Blender collaborate seamlessly, offering flexible options for creating and digitalizing 3D human models. Users can select MakeHuman for straightforward, parametric human modeling or MB-Lab, a powerful Blender add-on for advanced customization. Both tools integrate smoothly with Blender, allowing the use of conventional 3D modeling techniques alongside Blender's extensive features. This versatility ensures high-quality, detailed digital human models suitable for various applications in ergonomics, design, animation, simulation, and more. Exporting from MakeHuman to Blender involves setting the export scale units to meters and choosing between Collada (.dae) or Autodesk FBX formats—Collada being more reliable due to issues with FBX binary export. Collada files, though text-based and large, import directly into Blender, preserving textures saved in a "textures" folder. Ensure this folder accompanies the file during transfers to maintain texturing. Blender imports Collada files with adjustments needed for assets like eyes and hair. MakeHuman's support for OBJ and STL file formats also facilitates digital data transfer and 3D printing, with OBJ providing detailed texture and geometry storage and STL ensuring precise model printing. This dual support enhances versatility, enabling easy creation, refinement, and production of physical human models.

TABLE 8.11
MakeHuman: 3D Digital Design of Female

Tabs & Toolbar	Sub-tabs & Options	The Development Process of Completed 3D Digital Model of Female
Files: Load; Save; Export	Wireframe	
Modeling: Main; Body shapes; Gender; Face; Torso, Arms and legs, Random, Custom, Measure	Gender: Female; Age: 21; Muscle: 60.40%; Weight: 96.40%; Height: 168.40 cm	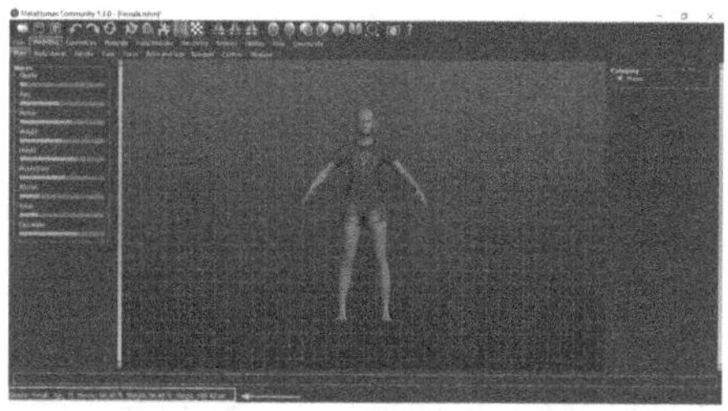
Geometries:	Clothes: Female casualsuit02;	
Geometries:	Eyes: High Poly;	
Geometries:	Hair: Long01;	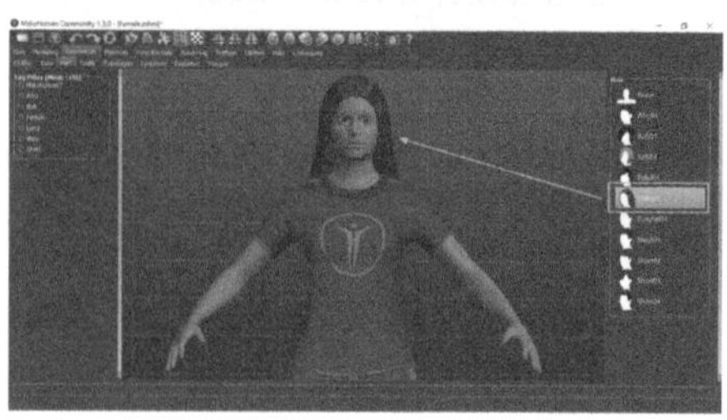

(*Continued*)

TABLE 8.11
(Continued)

Tabs & Toolbar	Sub-tabs & Options	The Development Process of Completed 3D Digital Model of Female
Geometries:	Teeth: Teeth Base;	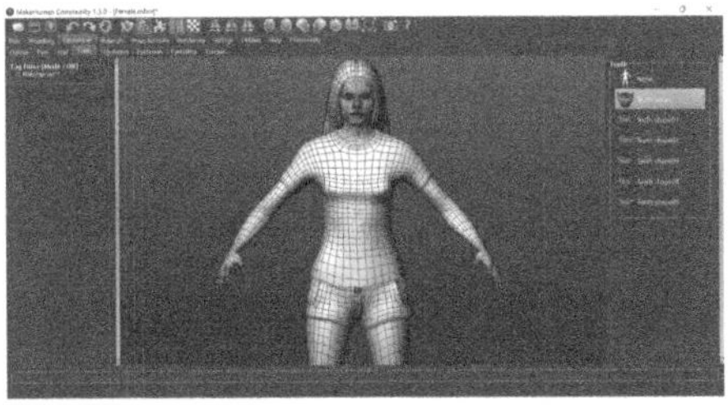
Geometries:	Eyebrows: Eyebrow01;	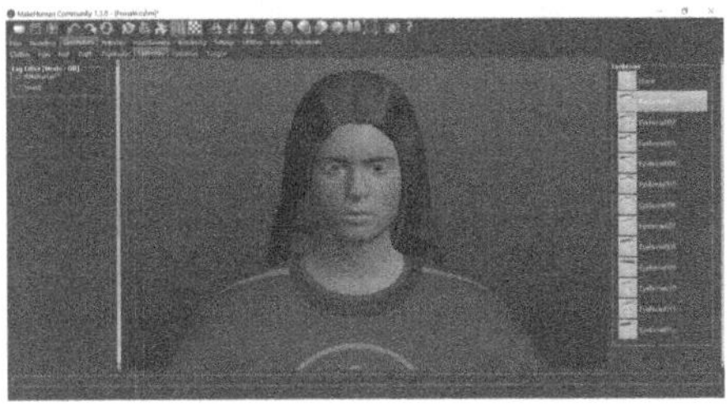
Geometries:	Eyelashes: Eyelashes04	
Completed 3D model view	Parameters finalization	
Pose/Animate	None (no armature)	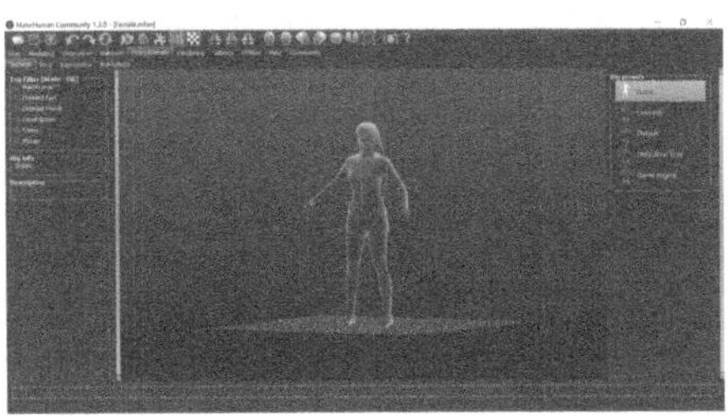

(*Continued*)

TABLE 8.11
(Continued)

Tabs & Toolbar	Sub-tabs & Options	The Development Process of Completed 3D Digital Model of Female
Pose/Animate	Default: MakeHuman default skeleton with detailed fingers, toes, and facial bones for advanced animations.	
Pose/Animate Skeleton & Pose: Standing04	Default: MakeHuman default skeleton with detailed fingers, toes, and facial bones for advanced animations.	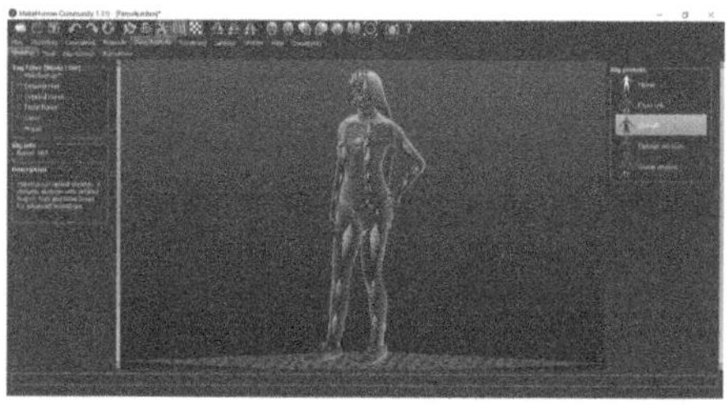
Pose/Animate Pose: Standing04	Default: MakeHuman default skeleton with detailed fingers, toes, and facial bones for advanced animations.	
Pose/Animate	Expressions: Smile01	
Pose/Animate	Expressions: Smile01	

TABLE 8.12
MakeHuman: 3D Digital Design of Male

Tabs & Toolbar	Sub-tabs & Options	The Development Process of Completed 3D Digital Model of Male
Files: Load; Save; Export	Wireframe	
Modeling: Main; Body shapes; Gender; Face; Torso, Arms and legs, Random, Custom, Measure	Gender: Male; Age: 25; Muscle: 73.90%; Weight: 122.00%; Height: 187.68 cm	
Geometries:	Clothes: Male casualsuit01;	
Geometries:	Eyes: High Poly;	
Geometries:	Hair: Long01;	

(Continued)

TABLE 8.12
(Continued)

Tabs & Toolbar	Sub-tabs & Options	The Development Process of Completed 3D Digital Model of Male
Geometries:	Teeth: Teeth Base;	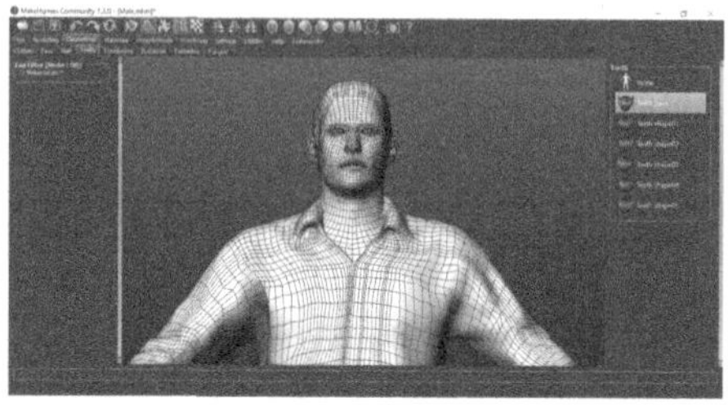
Geometries:	Eyebrows: Eyebrow12;	
Geometries:	Eyelashes: Eyelashes01	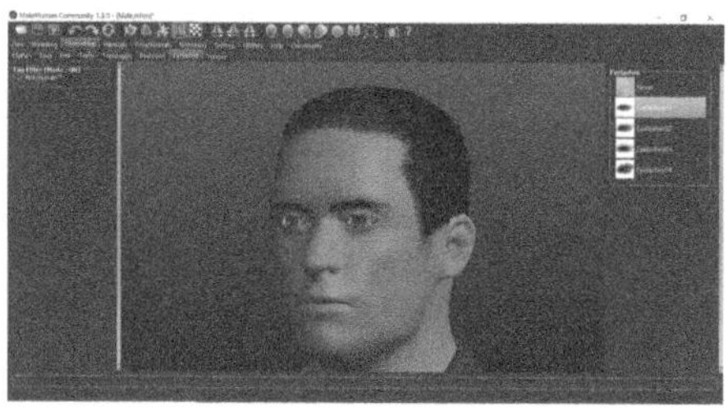
Completed 3D model view	Parameters finalization	
Pose/Animate	None (no armature)	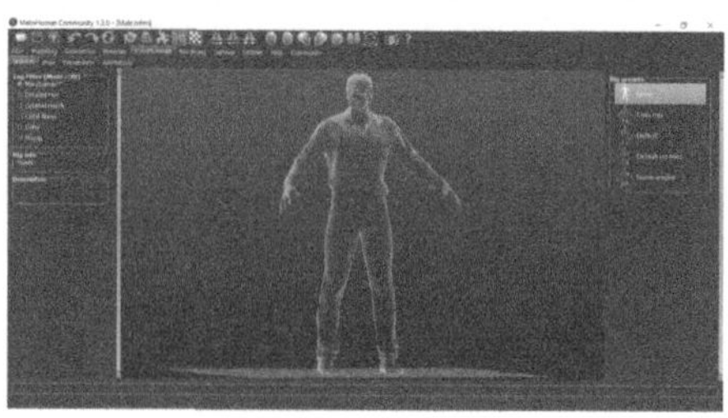

(Continued)

TABLE 8.12
(Continued)

Tabs & Toolbar	Sub-tabs & Options	The Development Process of Completed 3D Digital Model of Male
Pose/Animate	Default: MakeHuman default skeleton with detailed fingers, toes, and facial bones for advanced animations.	
Pose/Animate Skeleton & Pose: Run01	Default: MakeHuman default skeleton with detailed fingers, toes, and facial bones for advanced animations.	
Pose/Animate Pose: Run01	Default: MakeHuman default skeleton with detailed fingers, toes, and facial bones for advanced animations.	
Pose/Animate	Expressions: Vocalize:01	

REFERENCES

[1] Open Anatomy Project https://www.openanatomy.org
[2] OpenStax Anatomy and Physiology https://openstax.org/books/anatomy-and-physiology/pages/prefac
[3] Anatronica https://www.anatronica.com

[4] Dovramadjiev, Tihomir (2024). Exploring the dynamics of implantology: insights into anatomy and practical examples. *IETI Transactions on Ergonomics and Safety (TES)*, 2024, Volume 8, Issue 1, 37–48, USA, ISSN 2520-5439, DOI: 10.6722/TES.202404_8(1).0005. https://www.ieti.net/TES/2024V8I1/5.pdf
[5] MB-Lab Add-on: Blender Add-on https://blender-addons.org/mb-lab-add-on/
[6] MB-Lab Add-on: License https://mb-lab-docs.readthedocs.io/en/latest/license.html#license-of-the-database
[7] MB-Lab Add-on: Documentation https://mb-lab-community.github.io/MB-Lab.github.io/
[8] MB-Lab Add-on: Modeling Process https://mb-lab-docs.readthedocs.io/en/latest/model_process.html
[9] MB-Lab Add-on: Creation Tools https://mb-lab-docs.readthedocs.io/en/latest/creation_tools.html
[10] MB-Lab Add-on: Skin Editor https://mb-lab-docs.readthedocs.io/en/latest/skin_editor.html
[11] MB-Lab Add-on: After Creation Tools https://mb-lab-docs.readthedocs.io/en/latest/after_creation_tools.html
[12] MB-Lab: Save and export https://mb-lab-docs.readthedocs.io/en/latest/save_export.html
[13] MakeHuman: Version 1.1.1 https://static.makehumancommunity.org/makehuman/releases/releases_111.html
[14] MakeHuman: Version 1.3.0 https://static.makehumancommunity.org/makehuman/releases/releases_130.html
[15] MakeHuman Software The interface and its basic functions https://static.makehumancommunity.org/makehuman/docs/the_interface_and_its_basic_functions.html

9 3D Digital Practical Examples

Understanding 3D digital design is crucial for users and designers. For designers, it is important to master these state-of-the-art techniques. High-level design skills enable professional designers to apply advanced 3D modeling, motion capture (MoCap), and simulation tools effectively, ensuring that ergonomic solutions are both innovative and practical. This expertise contributes significantly to societal advancements by integrating cutting-edge technologies into ergonomic design.

9.1 ERGONOMICS AND MOTION CAPTURE (MOCAP)

3D digital design in ergonomics and human factors has significantly advanced through the integration of technologies like MoCap and various 3D digitalization techniques, enhancing our ability to analyze and improve human-machine-environment interactions. Initially, 3D software was primarily designed for internal system use, but current trends have expanded its application to interact with other systems and accommodate new file formats, thereby enabling more detailed 3D modeling that incorporates precise anthropometric data. Modern design technologies have bridged the gap between virtual and real 3D spaces through the use of 3D scanners, 3D printers, and other additive technologies. In the realm of ergonomics, MoCap technology has a very important role in creating simulated models of work processes within virtual environments, allowing for thorough examinations of how individuals engage with the human-environment-machine system. This amalgamation of MoCap and 3D computer design facilitates the analysis and optimization of ergonomic features, supports biomechanical analyses—such as evaluating athlete movements to develop personalized training methodologies—enriches animations in the computer game and 3D film industry by providing realistic character movements, and enables the creation of interactive presentations for compelling real-time demonstrations. Contemporary research and applications focus on utilizing open-source systems and freely available platforms to make these sophisticated technologies accessible to a broader range of users. This inclusive approach not only extends the reach of ergonomic studies but also enhances the practical deployment of these technologies across various domains, contributing to the development of more effective and user-friendly human-machine interfaces. The ongoing advancements and applications of MoCap and 3D digital design keep supporting of constructing a comprehensive database of three-dimensional digital models tailored to ergonomic needs. These developments were notably highlighted by the Bulgarian Association of Ergonomics and Human Factors (BAEHF) during the 34-Summer University on Ergonomics, where MoCap technology was presented and discussed, ultimately fostering a deeper understanding and implementation of these technologies for ergonomics and human factors needs.

DOI: 10.1201/9781003400929-9

MoCap technology, combined with specialized ergonomic software, is used for ergonomic design by enabling precise and accurate movement generation and real-time visualization of human activities.

- **Siemens Tecnomatix** MoCap **Toolkit [1]**. Integrating MoCap technology with human modeling significantly enhances design development and analysis by providing detailed tracking of full-body movements, including hands and fingers. This capability is important for improving ergonomic features in various applications, from industrial processes to product design, leading to more user-friendly and efficient systems. The Tecnomatix MoCap Toolkit exemplifies the transformative potential of MoCap technology, offering a sophisticated toolset for analyzing and optimizing human interactions in different environments. The toolkit helps prevent errors by allowing designers to observe how real humans interact with products or environments, identify issues through virtual reality immersion, and evaluate user experiences early in the design phase without physical prototypes. It also enables the rapid creation of full-motion scenarios and detailed analysis of physical stresses over time. By recording intricate postures and movements, the toolkit improves the review process of manufacturing and product designs. It supports multiple full-body tracking systems and specialized gloves for fine hand and finger motions. Real-time MoCap streams data to human simulation software, where digital avatars replicate movements accurately, adjusting for height, weight, and design criteria. This real-time interaction, enhanced by head-mounted displays (HMDs) for an immersive view, allows effective evaluation of visibility, reachability, and injury risks. This approach helps designers identify and correct usability issues without the cost of physical prototypes. The toolkit's ability to model complex human motions and behaviors is essential as design increasingly relies on accurate simulations, supported by motion tracking hardware and virtual reality capabilities. It enables rapid evaluation of design alternatives, assessment of complex postures, research of behavioral movement strategies, and evaluation of multi-person operations, offering a robust solution for ergonomic and human factors engineering in a PLM environment [2]. Table 9.1 shows examples of implementation of Siemens Tecnomatix software in practice: file Additional Resources [3, 4].
- **Dassault Systèmes DELMIA Ergonomics** is a premier 3D digital design software focused on ergonomics and human factors, offering a comprehensive suite of tools for designing, simulating, and validating human-centered manufacturing processes. It allows manufacturing engineers and designers to integrate ergonomic analysis directly within 3D virtual environments, facilitating the early detection and resolution of posturing issues, thereby enhancing workplace safety and efficiency. The software's advanced features, such as automatic posture prediction, ergonomic assessments, and in-depth analysis capabilities, enable users to visualize worker tasks in 3D and conduct detailed evaluations using virtual manikins with specific anthropometric and biomechanical attributes. DELMIA Ergonomics is packaged into specialized roles, including Ergonomic Workplace Designer, Ergonomics

Specialist, Ergonomist, and Work Safety Engineer, each tailored to meet the distinct needs of different organizational functions, providing targeted functionality and guidance. By simulating human interaction and ergonomic behavior from the initial stages of design, the software not only helps reduce the costs associated with work-related injuries but also enhances employee well-being and streamlines decision-making processes, even for those with limited ergonomic expertise, ultimately contributing to significant improvements in overall workplace design, operational efficiency, and safety [5].

- **ViveLab Ergo** is a lead ergonomic software designed to streamline workplace optimization through fast and precise ergonomic assessments, analysis, and planning. Catering primarily to engineers, EHS/HSE managers, and counselors in manufacturing companies, this software leverages innovative ergonomics measurement technology to evaluate and enhance employee movements, ensuring healthier workplace environments. ViveLab Ergo stands out with its built-in anthropometric database and seven comprehensive ergonomic analyses, which highlight unnecessary, time-consuming, and health-damaging movements caused by poor workplace design. By identifying these risk factors, ViveLab Ergo enables companies to make informed decisions that mitigate the adverse effects of forced postures and repetitive motions. The software's user-friendly interface and robust visualization capabilities facilitate detailed ergonomic evaluations, helping organizations improve their workspaces efficiently. This leads to a reduction in musculoskeletal disorders, enhanced employee well-being, and increased productivity, making ViveLab Ergo an invaluable tool for fostering safer and more ergonomic workplaces across various industries [6–8]. In the ViveLab Ergo system, seven types of analyses are implemented for evaluating human motion and posture: RULA (Rapid Upper Limb Assessment), OWAS (Ovako Working Posture Analysis System), NASA-OBI (NASA Task Load Index), ISO 11226 (Ergonomics – Evaluation of Static Working Postures), EN 1005-4 (Safety of Machinery – Human Physical Performance), Spaghetti Diagram (Motion Measurement), and Reachability Test. These analyses enable users to transfer human movements into anatomically correct and realistic models effortlessly. Additionally, ViveLab Ergo provides access to a large anthropometric databank, allowing for precise and comprehensive ergonomic assessments. The software's user-friendly interface ensures that ergonomic evaluations are accurate, enhancing workplace design and safety by leveraging extensive data and advanced modeling capabilities [9].

9.2 METHODOLOGY FOR INTERACTION BETWEEN MOTION CAPTURE (MOCAP) AND 3D SOFTWARE FOR ERGONOMICS NEEDS

To effectively develop high-quality simulations in a virtual environment for ergonomic and human factors applications, a comprehensive methodology must be employed that integrates MoCap systems with advanced 3D software. This process involves a combination of sophisticated technical expertise and careful planning to ensure accurate and useful results. The methodology begins with selecting the appropriate technology for capturing human movement, which may include using sensing equipment equipped with specialized sensors or relying on digital platforms with prerecorded movement databases. For optimal integration, it is crucial to have a deep understanding of three-dimensional geometry, the anatomical nuances of human bodies, and the dynamics of musculoskeletal movements. Accurate simulations necessitate the ability to interpret and manipulate complex motion data effectively. To facilitate the smooth interaction between MoCap systems and 3D software, it is vital to use the correct file formats that are compatible with both the MoCap system outputs and the 3D modeling tools. Key file formats include those that support detailed motion data and 3D geometry, ensuring that data is accurately represented and utilized within the simulation environment. Additionally, the methodology must include a structured approach to data processing and integration, ensuring that the captured motion data is seamlessly transferred and correctly interpreted by the 3D software. This approach is illustrated in Figure 9.1, which depicts an optimized workflow for achieving effective MoCap-3D software interaction tailored for ergonomic assessments. Table 9.2 provides a reference to important file formats, highlighting their role in ensuring compatibility and precision in the simulation process.

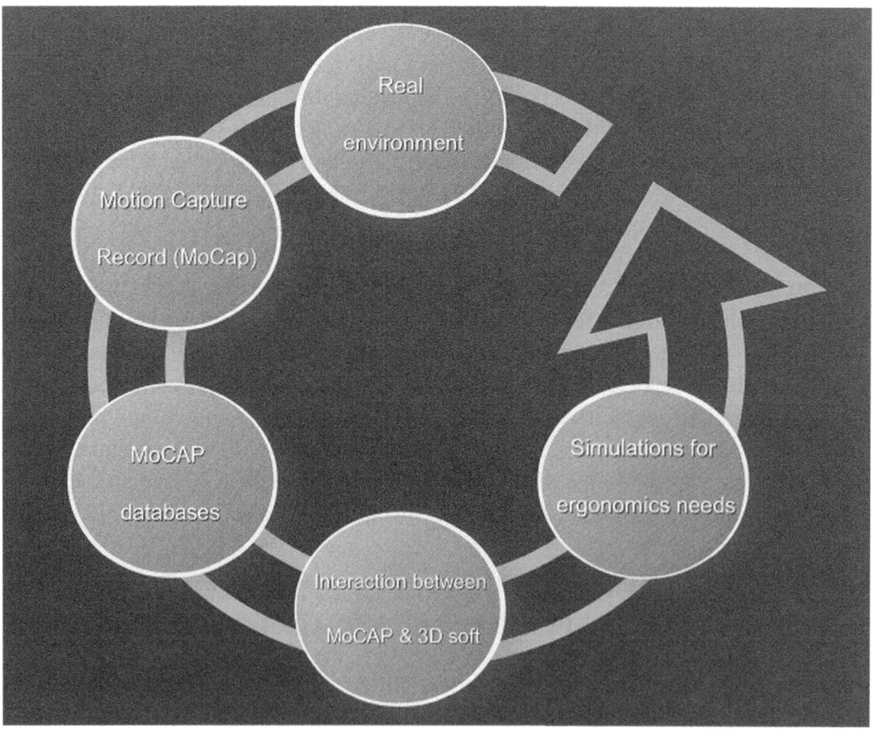

FIGURE 9.1 Methodology for interaction between MoCap and 3D software for ergonomics needs [10].

TABLE 9.2
3D Digital Design and MoCap Interaction [10]

Important file formats for 3D digital design and MoCap Interaction https://doi.org/10.6723/TERP.202112_5(2).0002 [10]

Full name	File format
Biovision Hierarchy/Motion Capture	*.bvh
Collada	*.dae
Filmbox	*.fbx
Stereolithography	*.stl
Wavefront	*.obj

The integration of MoCap technology with conventional 3D software represents a deep symbiosis that enhances both the accuracy and functionality of virtual simulations. This fusion begins with the capture of complex, nuanced movements from real-world subjects, which are digitized and stored in databases as MoCap data. These prerecorded motion profiles encompass a wide spectrum of activities, from routine actions to dynamic motor skills, and can be imported into sophisticated 3D software platforms. In these environments, the recorded motion data is not merely a static asset but serves as a foundational layer upon which further animations and interactions are developed. By leveraging tools such as Adobe Mixamo, ActorCore (AccuRig), Rokoko Studio, and others, designers can initially create 3D virtual human characters, either with or without skeletal structures (armatures), which are then subjected to MoCap-driven animations. This process allows for the simulation of highly realistic human movements within custom-designed 3D environments. Once the characters are animated, they can be seamlessly integrated into a virtual world where additional elements, including environmental features and objects, are introduced to test and refine ergonomic and human factor designs. This workflow not only enhances the precision of ergonomic assessments and user experience evaluations but also ensures that the digital simulations closely mirror real-world scenarios, thereby providing invaluable insights into how human factors influence interaction with designed systems and products.

The technical correlation between MoCap technology and conventional 3D software such as Blender, MakeHuman, and other platforms is immensely significant for a broad spectrum of designers and ergonomists, underscoring the importance of seamless digital data transfer and the strategic utilization of each tool's advancements to create comprehensive 3D digital human designs tailored to ergonomics and human factors. MoCap technology captures the subtle nuances of human movement through sophisticated sensors and translates these movements into detailed digital data, which serves as a critical foundation for animating 3D models. The ability to import this rich MoCap data into versatile 3D software like Blender allows designers to manipulate and refine animations with precision, leveraging Blender's extensive modeling, rigging, and animation features to enhance and adapt movements for specific ergonomic needs. Blender's capabilities in creating dynamic simulations and integrating environmental contexts enable a high degree of customization and realism in the digital representation of human interactions. Complementarily, MakeHuman provides an accessible interface for generating anatomically accurate and customizable 3D human characters, which can be rigged and animated using MoCap data, thereby creating a robust base for ergonomic simulations. The integration of MoCap with these tools facilitates a streamlined workflow where the accurate digital transfer of movement data into character models allows designers to conduct detailed analyses and simulations of human factors within virtual environments. This integration is vital for creating holistic and realistic 3D human designs that address ergonomic considerations effectively, ensuring that design solutions are based on precise and actionable data. By harnessing the strengths of MoCap technology alongside the advanced features of Blender and MakeHuman, designers and ergonomists can produce highly accurate simulations, leading to better-informed design decisions and enhanced user experiences across a range of applications.

- **Adobe Mixamo** stands out as one of the most popular and user-friendly tools for 3D rigging and animation, offering an exceptional platform for designers at all skill levels, from novices to professionals. Its seamless integration with Blender, Unity, MakeHuman, and others allows for effortless collaboration, providing a streamlined process for bringing 3D characters to life. Mixamo excels in delivering high-quality digital data transfer, ensuring that no information is lost during the process. Users can access a vast library of high-quality 3D characters that cater to a wide range of styles, including realistic, cartoon, fantasy, and sci-fi, each fully textured and rigged for immediate use in various creative projects. The automatic character rigging feature simplifies the process of uploading a custom character, automatically fitting a full human skeleton to the model, with customizable rigging options that can be optimized for mobile performance. Mixamo offers an extensive collection of motion-captured animations recorded from professional actors, which can be applied to any character. These animations are easily previewed and adjusted directly within Mixamo, giving users control over the final look and feel of each motion. Adobe Mixamo supports exporting characters and animations in multiple formats, ensuring compatibility with motion graphics, video games, films, and illustrations. Whether they need a single animation or a large batch, Mixamo's export optimizations help keep projects efficient and manageable [11].
- **ActorCore or AccuRig** represents evolution in free 3D digital design, particularly in the realms of ergonomics and human factors within character animation. This sophisticated platform offers an expansive library of 3D character animations that seamlessly integrates with all major 3D software programs, providing unparalleled versatility and efficiency. The robust capabilities of ActorCore are epitomized by its ability to retarget motions to selected characters effortlessly, a process simplified by comprehensive step-by-step guides and tutorials that cater to various programs. This user-centric approach is augmented by optimized Auto Setup and Import Presets, which streamline workflows and minimize manual adjustments. At the core

of ActorCore's innovation is its advanced rigging technique, development in transforming static images into dynamic 3D shapes by creating and manipulating connecting points that function as the "bones" of digital characters. This rigging technique not only enhances the realism and fluidity of character animations but also dramatically reduces production time by automating complex processes. ActorCore's emphasis on efficiency is further demonstrated by its support for A, T, and scan poses, accommodating a range of meshes from low to high resolution and even multiple meshes. Users can effortlessly export or upload their 3D tools and data directly to the ActorCore platform, which ensures access to an extensive library of production-ready assets. This direct integration with ActorCore's library significantly accelerates the animation process, enabling character artists to focus on creative aspects rather than technical hurdles. The platform's commitment to simplifying and automating the rigging and animation workflow makes it an indispensable tool for modern character artists striving to achieve both high-quality results and efficient production cycles. As of 2024, the software offers both free and paid services and 3D models, but the free version of AccuRig 1.0.3 is still available for download online [12].

9.3 EXAMPLES OF USING OF SPEN SOURCE SOFTWARE FOR RIGGING

Conventional human rigging: An exemplary model developed to illustrate these principles utilizes the software Blender in conjunction with the MB-Lab addon, forming the basis of a conventionally driven human character model. This process begins with the manual construction of the armature, or skeletal structure, which serves as the foundation for creating realistic movements and poses. To achieve this, rigging techniques are employed, where key positions and timeframes for each movement are meticulously defined, allowing for the accurate simulation of human dynamics. This method ensures that the character's motions align closely with real-world biomechanics, making it possible to evaluate and optimize ergonomic interactions. As the animation progresses, the system generates a dynamic human character that not only demonstrates functional movement but also integrates human factors considerations into its design. This model can be analyzed for its ergonomic effectiveness, assessing how well it accommodates and interacts with various environments, tools, or interfaces, thus providing valuable insights into the design of products and systems that better align with human capabilities and limitations. Figure 9.2 shows a conventional human rigging result developed in an open tutorial by Dovramadjiev T. using Blender software. Rigging and animation of 3D human model (2019) [13], as presented in the scientific article: Tihomir Dovramadjiev (2021), MoCap and 3D computer design for ergonomics needs. IETI TERP 2021 VOLUME 5 ISSUE 2, Pages: 8–17, DOI: 10.6723/TERP.202112_5(2).0002 [10].

- **3D-driven human characters**: To design 3D-driven human characters for ergonomic applications using open-source software and MoCap databases, a systematic approach is essential, encompassing precise anatomical and

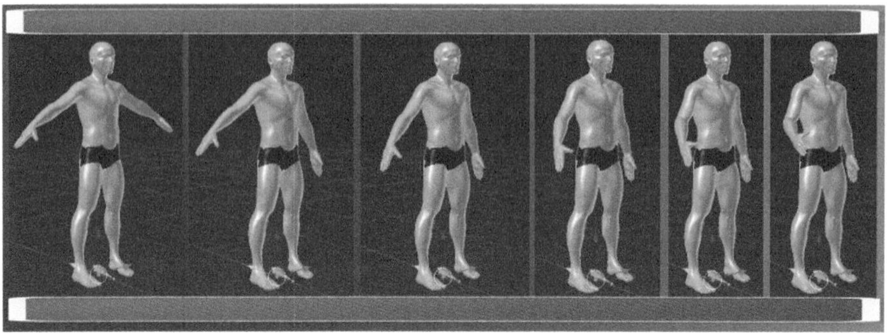

FIGURE 9.2 Blender + MB-Lab: conventional human rigging [10, 13].

contextual parameters to ensure realistic and functional results. This involves establishing the fundamental characteristics of the character, including, but not limited to, gender, age, muscle distribution, height, weight, and any other relevant physical attributes. For instance, when tasked with creating a 3D model of a working man with the specified physical parameters—age 41, muscle mass at 76.20%, weight at 119.80 kg, and height at 184.92 cm—the process begins by defining these attributes comprehensively. The creation of the 3D model starts with the accurate parametrization of these features, resulting in a digital representation that matches the given specifications. Once the model is constructed, an armature—a skeletal structure based on MoCap data—is built and integrated into the model to facilitate realistic movement and interaction. This armature, as depicted in the process flow, allows the model to mimic real-life actions and postures (Table 9.3). The final step involves exporting the model and its armature in a compatible file format, such as *.fbx, to ensure seamless integration with various 3D platforms for simulation and further ergonomic analysis. This methodology not only

TABLE 9.3
Motion Capture (MoCap) and 3D Digital Design [10]

Software / Platform:	Motion Capture (MoCap) and 3D digital design. Example source: https://www.ieti.net/TERP/2021V5N2/2.pdf [10]
Interface of MakeHuman software. Adding Motion Capture data from CMU Graphics Lab Motion Capture Database directly in MakeHuman environment (right image) http://mocap.cs.cmu.edu/ [14].	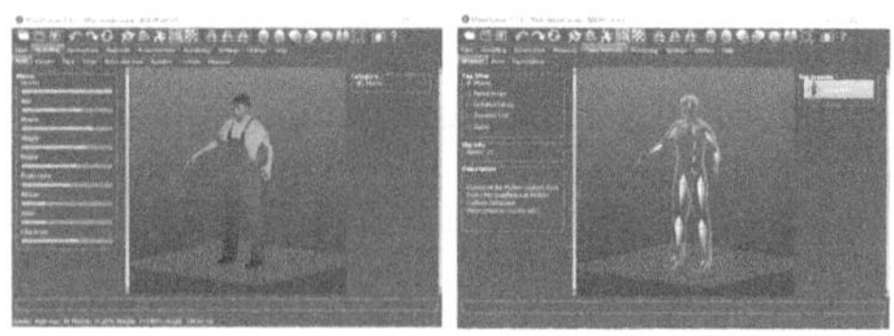 3D developed model: (left image) 3D worker physical parameters: Age: 41, Muscle: 76.20%, Weight: 119.80%, Height: 184.92 cm.

(Continued)

TABLE 9.3
(Continued)

Software / Platform:	**Motion Capture (MoCap) and 3D digital design. Example source: https://www.ieti.net/TERP/2021V5N2/2.pdf [10]**
Interface of Blender software. Imported from MakeHuman 3D worker model.	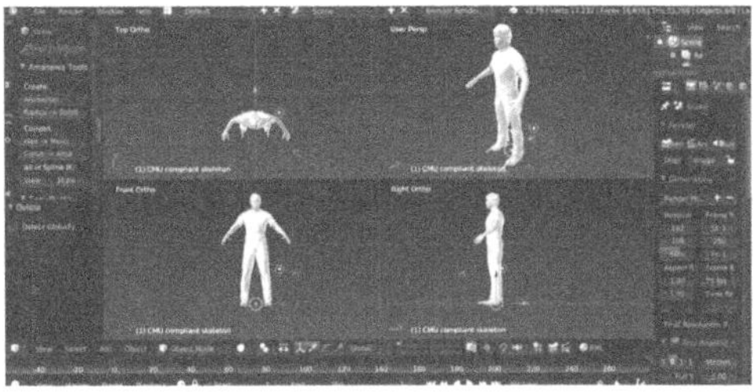 Blender software: Check the specific developed human model, which contains: Vertices: 17,232; Faces: 16,663
Adobe Mixamo platform. 3D model of a male worker is transferred to the Adobe Mixamo platform, where the necessary actions are applied for recorded movements of work processes. The figures show selected frames from the moving working model: 82, 97, 129, 157, and 234.	 Video presentations of completed versions of a digital dynamic model of a human character performing certain work processes are available at the following links: • Push Start https://www.youtube.com/watch?v=Zu4rtgn27Hc • Putting down https://www.youtube.com/watch?v=i9QK4JXiufY • Bartending https://www.youtube.com/watch?v=0CRVhMNj2E8
P3D online platform* The P3D online platform is detailed present in Section X. Additional digital resources 3.3D digital interactive presentation. The development of the 3D model assignment resulted in a high-quality and well-constructed mesh. This example contributes to building a digital library of three-dimensional models, which can be integrated into the work environment and used for ergonomic research.	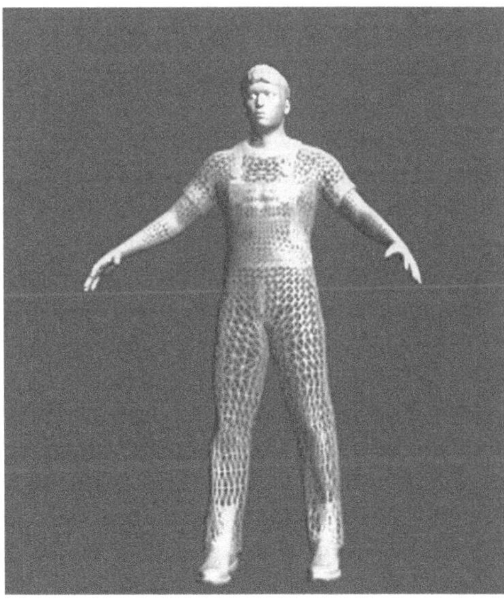 Real-time presentation of the 3D worker model and mesh view (X-rey + wireframe; other) [15].

ensures the accuracy of the ergonomic simulation but also enables a detailed assessment of how the character interacts with their environment, thereby contributing to more effective design and evaluation of ergonomic solutions.

9.4 ERGONOMICS AND BIOENGINEERING FOR INCLUSIVE HEALTHCARE (IMPLANTS, PROSTHESIS, MASKS)

In healthcare, the integration of ergonomics and bioengineering through 3D digital design is very important for developing protective products like implants, prostheses, and masks. This approach leverages advanced scanning technologies to capture precise anatomical details, allowing for the design of highly customized, ergonomic devices that ensure optimal fit and function. 3D design software facilitates the creation of implants and prosthetics that seamlessly integrate with the human body, enhancing comfort and biomechanical performance. For protective equipment such as masks, ergonomics ensures that designs are comfortable and effective, while bioengineering contributes to the selection of materials that enhance durability and functionality. Virtual simulations and digital prototyping play a crucial role in rapidly testing and refining these designs, particularly in response to public health needs. By utilizing additive manufacturing technologies like 3D printing and CNC machining, these protective products are produced with high precision, improving patient outcomes and ensuring that medical devices meet both ergonomic and functional requirements. This synergy between ergonomics, bioengineering, and 3D digital design enhances the development of tailored healthcare solutions that promote safety, comfort, and effectiveness in various medical applications.

- **WASP MED Add-on for Blender 2.8** was launched on July 4, 2019, the WASP MED Add-on is an open-source 3D software tool designed for the prosthetics and orthotics market, projected to reach USD 12.28 billion by 2025. Developed by WASP MED in collaboration with Prof. Alessandro Zomparelli, it bridges the gap between expensive professional software and versatile open-source tools. Integrated with Blender, it simplifies creating 3D-printable prosthetic and orthotic designs from 3D scans with commands for fixing, cropping, and exporting meshes. The tool promotes accessibility and innovation in medical devices [16]. The open-source tool WASP MED Add-on, compatible with Blender versions 2.8, 2.82, and 2.92, enhances the quality and efficiency of medical device design by offering a comprehensive set of features. It supports precise modeling from 3D scans with tools for fixing, correcting, cropping, and refining shapes, which is crucial for creating custom-fit devices. The add-on streamlines the design process with step-by-step commands, making it easier to produce complex, 3D-printable models that are accurate and reproducible. By integrating these specialized functions with Blender's native capabilities, the add-on improves the production of prosthetics and masks, ensuring comfort and functionality. Its ability to manage thicknesses and borders precisely, combined with support for 3D printing, bridges the gap between costly professional software and accessible open-source solutions. This fosters innovation and collaboration within the 3D design and healthcare communities, promoting the development of high-quality, affordable medical devices globally [17].

- **Blender 2.82**, released shortly after 2.81, brings a host of significant updates and new features. Notably, the modeling workflow has been enhanced with Tool Drag Actions, allowing users to quickly switch tools by dragging in the viewport, thereby streamlining the process of selecting and transforming objects. The introduction of custom bevel profiles enables the creation of complex bevel shapes using custom curves, a highly anticipated feature for hard surface modeling. The new Weld Modifier automates the merging of close vertices, akin to the "Merge by Distance" function, but as a nondestructive modifier. The Solidify Modifier also received a major upgrade with a "Complex" mode, which better handles intricate geometry by producing even thickness and flat faces, particularly benefiting 3D printing enthusiasts. These updates demonstrate Blender's rapid development pace and its commitment to improving user experience and tool functionality [18].
- **Blender 2.92**, the first release of 2021, introduces a range of exciting new features and improvements across various aspects of the software. One of the most notable additions is the Geometry Nodes, which enable procedural modeling through a node-based system, enhancing flexibility and control. The APIC fluid simulation method offers more realistic fluid dynamics, particularly for smaller-scale scenes. In rendering, Eevee now supports Cryptomatte and Arbitrary Output Variables (AOVs), while Cycles has received significant speed optimizations. Sculpting tools have been expanded with new capabilities like the Elastic Snake Hook brush and improved Face Set editing options. The Grease Pencil toolset has been enhanced with Curve Editing mode and better interpolation algorithms. The user interface has seen refinements such as improved Outliner functionality and more intuitive modifier management. Other areas like animation, rigging, and video editing have also received meaningful updates, making Blender 2.92 a robust and versatile tool for 3D artists [19]. The file: Additional Resources: Tables 9.4 and 9.5 provide information for download parameters.
- **My Face Mask Add-on** is an open-source tool specifically designed for creating 3D-printable masks from 3D scans. This add-on simplifies the modeling process through a guided step-by-step workflow, while also allowing users to leverage the full suite of Blender's native tools for further customization. It includes features such as scene setup, import and alignment of scans, and remeshing to fix common scan issues. Users can easily trace mask borders, adapt the mask to the face, and adjust specific areas like the filter placement and mesh control points. It provides options to set dimension variables like nose pressure, thickness, and offset for optimal fit and printing compatibility. The add-on also prepares the model for 3D printing, highlighting potential issues and enabling users to tag the mask for identification. The final model can be exported in STL format, ensuring a streamlined and efficient design process for producing high-quality,

custom-fit masks suitable for various healthcare applications [20]. To install the My Face Mask Add-on for Blender 2.82, first download the add-on from GitHub (https://github.com/alessandro-zomparelli/myfacemask). Open Blender and navigate to Edit > Preferences > Add-ons. Click "Install from file" and locate the downloaded .zip file. Once installed, enable the My Face Mask Add-on by checking the corresponding box in the add-ons tab. The add-on is now ready to use and can be accessed on the right side of the viewport panel. Importing and aligning 3D scans (.obj or .stl files) is simplified, with automatic interface adjustments and alignment to global reference axes. The remesh function allows for fixing scan errors and selecting the desired resolution based on computational capacity. Users then trace the mask borders using a selection brush, ensuring proper fit around the nose, cheeks, and chin. The "Adapt mask" feature aligns the mask model to the face, with options for symmetric borders. Adjustments for filter alignment and manual editing ensure a precise fit, while "Show Holes" visualizes and places holes for elastic ropes. Dimension variables like "Nose Pressure," "Thickness," and "Offset" further customize the mask's fit and compatibility with specific 3D printing parameters. The add-on prepares the model for printing, allowing for tag imprints and adjustments to avoid printing issues. Finally, the model is exported as an STL file, ready for 3D printing, ensuring a custom and precise fit for each individual. For the convenience of users and designers, WASP has presented video tutorials, which are presented in the file: Additional Resources Table 9.6. The process of installation of 3D WASPs addons: Med Add-on and MY FACE MASK Add-on to Blender is shown in Table 9.7.

TABLE 9.7
3D WASP: Install MED and My Face Mask Add-ons to Blender

Steps	Install 3D WASP Add-ons to Blender	Blender Software Interface
1	Start Blender software > File > User Preferences > Add-ons > Install Add-on from file > MED file	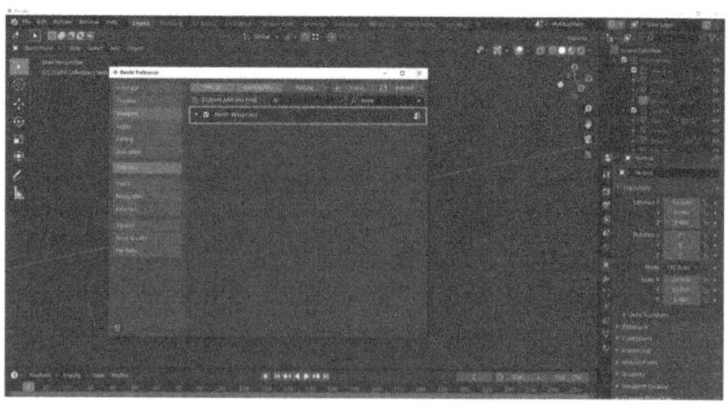

(Continued)

TABLE 9.7
(Continued)

Steps	Install 3D WASP Add-ons to Blender	Blender Software Interface
2	The 3D WASP Med Add-on position (Blender software 3D environment)	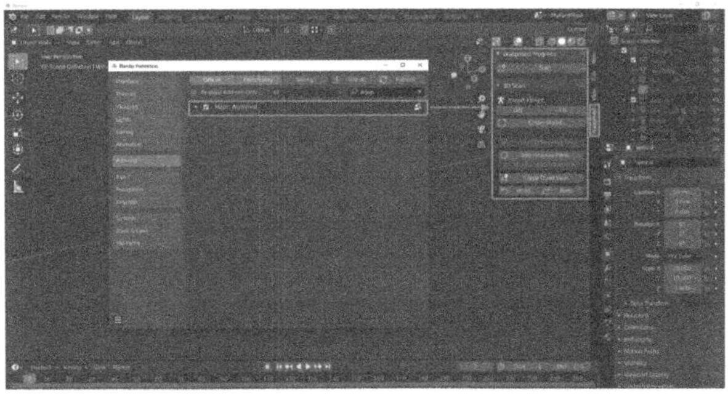
1	Start Blender software > File > User Preferences > Add-ons > Install Add-on from file > My Face Mask file	
2	The 3D WASP My Face Mask Add-on position (Blender software 3D environment)	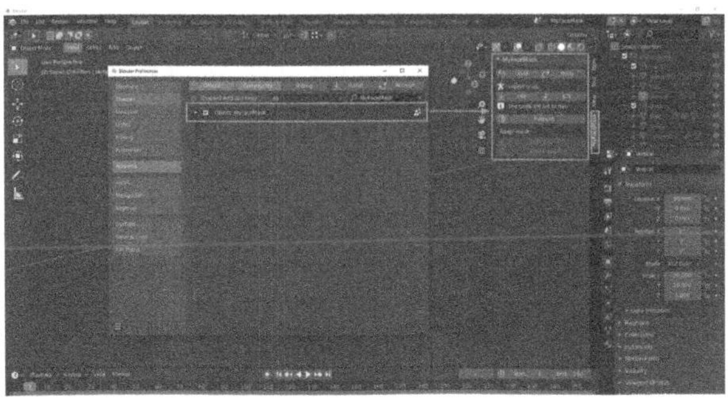

Dentistry Add-ons: Open Dental CAD – ODC & Dents 3D – Human Teeth Addon (+Nikitron tools + related Add-ons).

The field of 3D digital design in dental healthcare involves creating precise and biocompatible dental models using advanced technologies. This domain merges the principles of ergonomics and human factors to ensure that dental devices not only fit comfortably and function effectively but also align with the physiological and psychological needs of patients. Ergonomics in this context addresses the design and arrangement of equipment to enhance the dentist's performance and patient comfort, while human factors consider the interaction between the dental equipment and users, focusing on usability, safety, and efficiency. The integration of open-source software like Blender and FreeCAD has improved the creation of 3D dental models, offering cost-effective and accessible tools for designing and simulating dental structures such as crowns, bridges, and implants [21]. These software platforms facilitate the generation of precise 3D geometries through conventional modeling techniques, direct scripting, and specialized add-ons like "Dents 3D – Human Teeth Addon" [22] and "ODC – Open Dental CAD" [23] (Figure 9.3 Human Teeth, Figure 9.4 Dental Implants, Figure 9.5 Dental crowns).

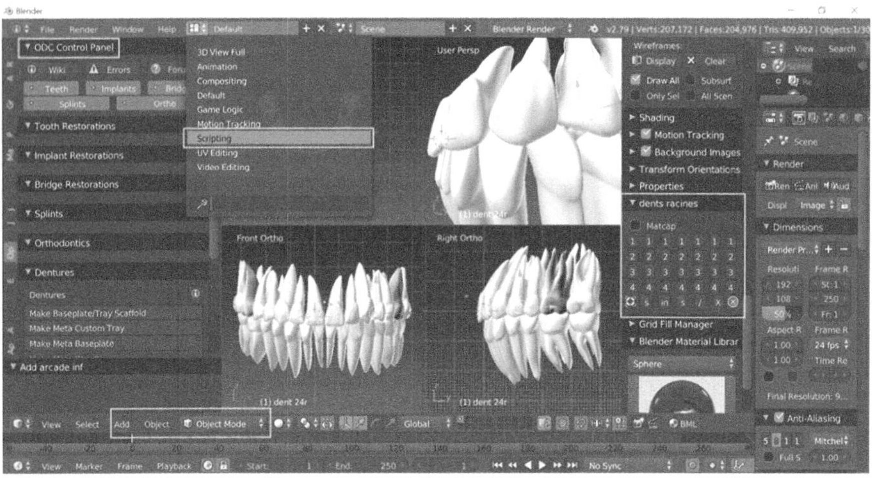

FIGURE 9.3 Interface (Theme Science Lab) of Blender software (version 2.79b) and basic possibilities for creating 3D dental models. In the green boxes are marked: the specialized dental add-ons: "ODC – Open Dental CAD" and "Dents 3D – Human Teeth Addon", the operating mode "Scripting," and the main options for conventional creation of 3D geometry. (In the middle are presented rows of upper and lower jaw obtained through Dents 3D addon) [21].

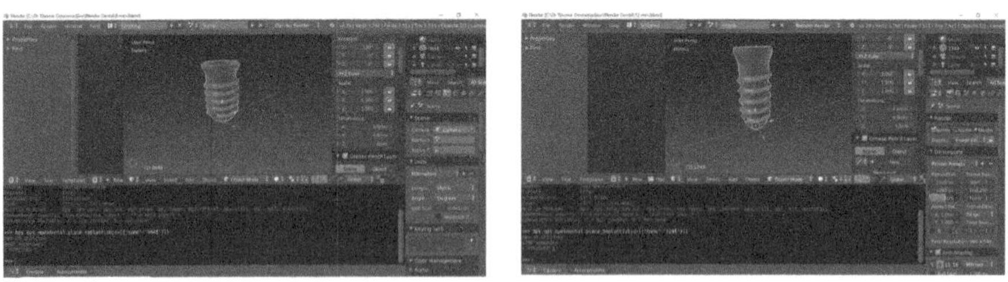

FIGURE 9.4 Standardized dental implants built using Python scripts, via ODC add-on in Blender software environment (left) dental implant 0848; (right) dental implant 1248 [21].

The Python scripts for them are: 4.8 mm/4.8 mm / 8 mm (left): bpy.ops.opendental.place_impl ant(objs=[{"name":"0848"}]); 4.8 mm/4.8 mm / 12 mm (right) bpy.ops.opendental.place_imp lant(objs=[{"name":"1248"}])

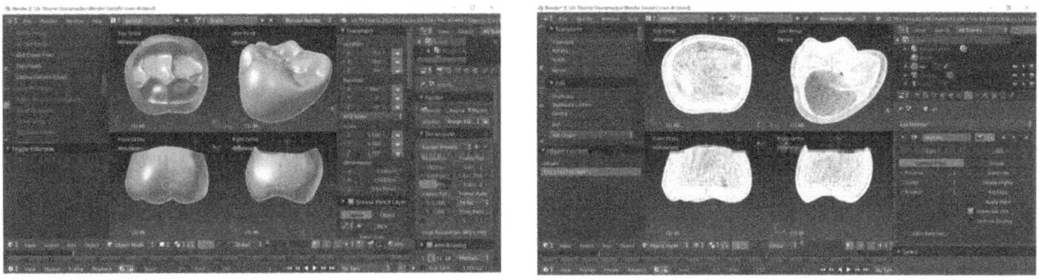

FIGURE 9.5 ODC standardized dental crown with N: 46, 3D created using Python scripts in the environment of Blender software (left) 3D basic model of the crown (right) 3D ready model of the crown with a thickness of 0.5 mm [21].

The Python script is as follows: ODC Dental Crown: bpy.ops.opendental. get_crown_form (objs=[{"name":"46"}])

The process involves obtaining data from clinical cases, digitizing problem areas using 3D scanners, and then creating digital models. Critical to this process is the calculation of the volume and mass properties of materials such as gold (Au), zirconium (Zr), titanium (Ti), and the titanium alloy Ti6Al4V. These materials are chosen for their biocompatibility, crucial for preventing adverse reactions in patients and ensuring the long-term success of dental restorations. The use of these advanced modeling techniques not only improves the precision of dental implants (Figure 9.4) and restorations but also plays a significant role in optimizing the design process, reducing errors, and enhancing patient outcomes. The application of mathematical calculations within these software environments, facilitated by add-ons like Nikitron tools (https://github.com/nortikin/nikitron_tools) for Blender and Macro FCInfo for FreeCAD (https://wiki.freecadweb.org/Macro_FCInfo), allows for accurate determination of the material mass, which is crucial for budgeting and financial planning. Using these add-ons, the mathematical values for the volume of the three-dimensional geometry are determined and then transferred to specialized online calculators, such as "SensorsOne" (https://www.sensorsone.com/volume-and-density-to-mass-calculator/). To accurately calculate the mass of the material used, it is necessary to define it with standardized data for the specific material (https://www.theworldmaterial.com/density-of-metals/ and https://ntrs.nasa.gov/api/citations/19720022814/downloads/19720022814.pdf). This financial aspect is important given the high cost of materials like gold, which can significantly impact the overall expenses in dental care. By accurately predicting the amount of material needed, these technologies help in cost estimation and management, thereby making high-quality dental care more accessible. Accurate calculation of the financial impact in the dental industry becomes feasible once the mass of the material, such as gold used in dental crowns and bridges, is determined. Specialized calculators like "ARCH EnterPrices: SILVER AND GOLD VALUE CALCULATOR" (https://www.archenterprises.com/silver-gold-calculator.html) and GoldPriceZ (http://goldpricez.com/calculator/gold-rates) enable precise financial value computations, leveraging mathematical data and real-time pricing. This calculation is crucial in the management and production processes within the dental sector, which involves a wide array of activities related to the global production of dental accessories, tools, components, and services. The financial scale of these activities underscores the importance of having accurate data on material costs, which facilitates meticulous financial planning. By incorporating the cost of expensive materials like gold, alongside production, labor, and distribution expenses, dental companies, clinics, and laboratories can optimize their operations. This comprehensive financial oversight not only aids in efficient resource management but also ensures the setting of accurate prices for end users, thereby contributing to a transparent and sustainable dental healthcare system globally. The ability to precisely manage these costs is a key factor for business success in the dental sector, allowing for the delivery of high-quality services and products while maintaining profitability and competitiveness (Figure 9.6).

The full information on mathematical and finance calculations is mentioned in an article by Tihomir Dovramadjiev, titled "Application of open source software for creating three-dimensional dental models, mathematical calculating the mass properties of gold, zirconium and titanium and financial impact on the health care system" ieti.net/TERP/2022V6N1/IETI%C2%A0TERP%C2%A0V6%C2%A0I1%C2%A015-28.pdf published in the IETI Transactions on Engineering Research and Practice, provides a comprehensive and transparent view of the study's findings. It includes detailed mathematical calculations and financial impact assessments related to the development of 3D digital dental models. The data covers the exact volume and mass of dental implants and crowns made from materials such as gold, zirconium, and titanium, including the Ti6Al4V alloy, with up-to-date gold pricing information. The article thoroughly documents the accuracy of the 3D models generated using open-source software, such as Blender, and verifies that there are no inaccuracies in the polygonal mesh or the thickness of the 3D geometry. It also describes the methodology used for generating these models and their functional readiness for

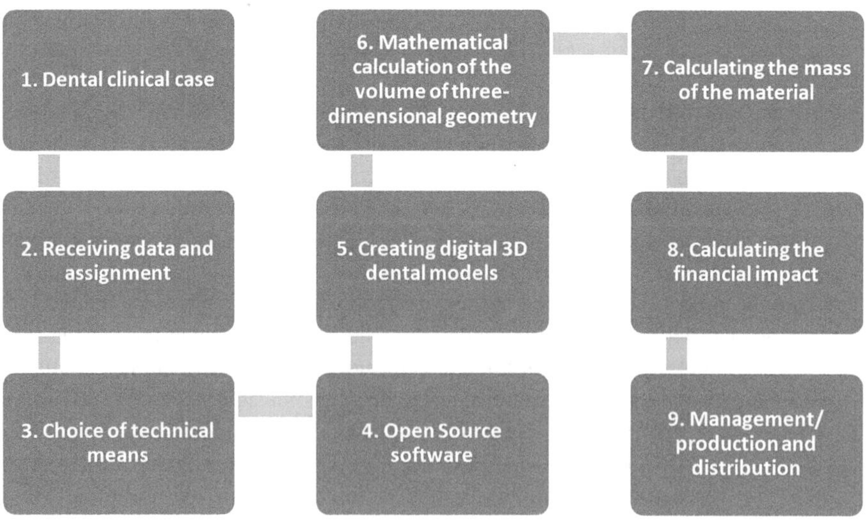

FIGURE 9.6 Optimized methodology for creating three-dimensional digital geometry of dental models and financial impact, https://doi.org/10.6723/TERP.202207_6(1).0002 [21].

production. This extensive data and analysis are fully accessible and open for reading on the IETI website, allowing anyone interested to examine the detailed calculations, financial implications, and results of the study in their entirety.

REFERENCES

[1] Siemens Tecnomatix Motion Capture Toolkit https://www.plm.automation.siemens.com/global/en/products/manufacturing-planning/human-centered-design-simulation.html

[2] 4DSYSCO: Motion Capture Toolkit for Tecnomatix human applications https://4dsysco.com/pdf/Siemens/Motion%20Capture%20Toolkit%20For%20Tecnomatix%20Human%20Applications.pdf

[3] Siemens: Human-centered design and simulation https://plm.sw.siemens.com/en-US/tecnomatix/human-centered-design-simulation/

[4] Siemens: Virtual reality and human simulation at GM https://blogs.sw.siemens.com/tecnomatix/virtual-reality-and-human-simulation-at-gm/?pk_vid=f42c1f20453ac57e1721811495a4968e

[5] Dassault Systèmes Delmia: Design Ergonomic Workstations Virtually https://www.3ds.com/products/delmia/industrial-engineering/ergonomics

[6] ViveLab Ergo: Ergonomic-risk assessment software https://www.vivelab.cloud/

[7] ViveLab Ergo: VIVELAB ERGO FREE DESKTOP VERSION https://www.vivelab.cloud/free-demo-desktop

[8] ViveLab Ergo: Download https://drive.google.com/file/d/14pjJZr3HlP0g9qBTICtAvMCIti4W3_8O/view

[9] ViveLab Ergo: Guide https://drive.google.com/file/d/1-3Eh66SImL3hlLMgBxkWpYF457uWaO37/view

[10] Doramadjiev, Tihomir (2021). Motion capture (MoCAP) and 3D computer design for ergonomics needs. *IETI TERP 2021* Volume 5 Issue 2, 8–17, https://doi.org/10.6723/TERP.202112_5(2).0002, https://www.ieti.net/TERP/2021V5I2.html, https://www.ieti.net/TERP/2021V5N2/2.pdf

[11] Adobe Mixamo https://www.mixamo.com

[12] ActorCore or AccurRig https://www.softpedia.com/get/Multimedia/Graphic/Graphic-Editors/AccuRIG.shtml

[13] Doramadjiev, Tihomir (2019). Blender software. Rigging & Animation of 3D Human Model. Zenodo. https://doi.org/10.5281/zenodo.13337895

[14] CMU Graphics Lab Motion Capture Database http://mocap.cs.cmu.edu/

[15] Doramadjiev, Tihomir. Real-time presentation of the 3D worker model and mesh view (X-Rey + Wireframe) https://p3d.in/P6B87/wire

[16] 3DWASP: The open-source 3D software for P&O now exists https://www.3dwasp.com/en/wasp-med-add-on-blender-2-8/

[17] 3DWASP: WASP MED developed an Add-on for Blender 2.8 https://www.3dwasp.com/en/addon-blender-wasp-med/

[18] Blender Nation: What's new in Blender 2.82? All the major changes and updates https://www.blendernation.com/2020/03/12/whats-new-in-blender-2-82-all-the-major-changes-and-updates-promoted/

[19] CGCookie: What's new in Blender 2.92? https://cgcookie.com/posts/everything-new-in-blender-2-92-modeling-sculpting-ui

[20] 3DWASP: MY FACE MASK Add-on Blender 2.82 https://www.3dwasp.com/en/my-face-mask-addon-blender/

[21] Doramadjiev, Tihomir. (2022). Application of open source software for creating three-dimensional dental models, mathematical calculating the mass properties of gold, zirconium and titanium and financial impact on the health care system. *IETI TERP*, Volume 6, Issue 1, 15–28, https://doi.org/10.6723/TERP.202207_6(1).0002
[22] Dents 3D – Human Teeth Addon http://byaapplication3d.blogspot.com/p/ents-3d.html
[23] Open Dental CAD – ODC. Dental Design Tools for Blender https://sites.google.com/site/blenderdental/

10 Additional Digital Resources

By incorporating add-ons and extensions, designers can refine their workflows, ensuring that their models are not only aesthetically pleasing but also ergonomically sound and user-friendly. The ability to share and integrate these add-ons within the community allows for the open exchange of experiences and best practices, which, in turn, drives innovation and continuous improvement in 3D design processes. Thus, add-ons are essential for advancing digital design practices, providing the tools necessary to create designs that are both functional and aligned with human factors principles.

10.1 DIGITAL SOFTWARE USEFUL EXTENSIONS (ADD-ONS)

The Human Generator Add-On for Blender 3D is an advanced and user-friendly tool designed to create realistic human characters with extensive customization options. The add-on allows users to choose between male and female base models, which can then be customized in the creation phase with settings for body type, height, and facial features, ensuring each character's physique is unique. Skin settings offer adjustments for tone, redness, saturation, and more, while hair options include various styles and lengths for both head and facial hair. Once the creation phase is complete, the detail phase allows for continuous editing, including posing, dressing, and facial expressions, with features like automatic rigging and posing enhancing usability. The add-on also supports bulk character generation using markers, making it ideal for populating scenes with varied human models. Although the randomization of bulk-created characters could be improved, creating smaller batches can yield better diversity. Additional content packs for hair and clothing can be purchased or customized, further expanding creative possibilities. Overall, the Human Generator Add-On is a powerful asset for Blender users seeking to create realistic human characters efficiently [1]. Users can choose a gender and select from available starting human presets, filtering through categories for specificity. After selecting a preset, "Generate new human" creates the character at the 3D cursor's location, set with shift + right-click. If the creation panel isn't visible, ensure no human is selected by choosing another object or using the deselect button. The Python API offers additional customization for those familiar with coding. Once created, the section overview panel allows for detailed edits, including body proportions, facial features, height, skin tone, hairstyles, clothing, and footwear, with support for posing and facial expressions using Rigify rig and motion capture [2, 3].

The **Procedural Crowds add-on for Blender**, developed by Difffuse Studio, is a powerful tool for generating realistic and dynamic 3D crowds instantly. It offers a wide range of customization options, enabling users to tailor crowd behavior and

DOI: 10.1201/9781003400929-10

arrangement to fit their specific scene needs. The add-on includes a library of 3D human characters that can be animated to idle, walk randomly, follow curves, or snap to surfaces while avoiding collisions. Users can create circular gatherings, position crowds in stadium seating with options to switch between idle and cheering states, and even animate crowds to follow predefined paths. The add-on is versatile, allowing the integration of user-provided assets alongside the included characters, which can be manually placed and animated with various actions like cheering or walking. With controls accessible through the sidebar, the add-on offers six main crowd configuration options, each designed to adapt to different scene requirements, making it an efficient and user-friendly solution for creating detailed and lifelike crowd simulations in Blender [4].

10.2 DIGITAL HUMAN CALCULATORS: ERGONOMICS, HEIGHT PERCENTILE, BODY MASS INDEX (BMI), AND OTHER

Digital human calculators are essential tools in 3D digital design for ergonomics and human factors, allowing designers to create personalized and ergonomic solutions. These calculators use advanced algorithms and extensive databases to model human anthropometry, height percentile, body mass index (BMI), and other key metrics. By inputting demographic data like age, gender, and ethnicity, designers can generate accurate 3D models that reflect the anatomical dimensions of diverse user groups, ensuring inclusivity in design. The integration of mathematical anthropometric values is crucial for enhancing the quality of 3D digital designs, making virtual environments more ergonomic and user-friendly. These tools also assess posture, reach, and movement, optimizing workplace layouts and product designs. Prominent tools like **OHCOW's Anthropometric Calculator** and **ErgoFellow 3.0** software stand out for their comprehensive capabilities [5, 6].

The UPLIFTDESK ergonomics calculator is a versatile tool designed to swiftly determine optimal settings for sit-stand desks, ergonomic chairs, monitor arms, and keyboard trays based on individual height (https://www.upliftdesk.com/ergonomic-calculator/). By factoring in shoe height and other relevant adjustments, such as clothing allowances, the calculator ensures precise ergonomic measurements. It integrates insights from the 1988 Anthropometric Survey of United States Army Personnel to provide reliable data suitable for both male and female body dimensions. The tool emphasizes the importance of ergonomic principles, encouraging users to adjust their workstations to fit their unique postures and preferences throughout the day. This approach not only enhances comfort and reduces strain but also promotes productivity by optimizing ergonomic conditions tailored to individual needs (Table 10.1) [7]. The table shows the results obtained for entered human heights of 152 to 193 cm.

A similar application that is freely available is the online calculator: **Desk Height Calculator Widget** by Inch Calculator [8]. Integrate the Desk Height Calculator into the website effortlessly, compatible with a variety of website builders, including WordPress. Choose a widget style that suits to site, then simply copy the HTML code provided and paste it into the web page editor. For WordPress users, ensure to activate the "code editor" to accurately place the widget as needed. Platforms like

TABLE 10.1
Ergonomic Calculations https://www.upliftdesk.com/ergonomic-calculator/ [7]

		Sitting Position		Standing Position	
cm	Seat Height Range (cm)	Sitting Eye Height Range (cm)	Sitting Elbow Height Range (cm)	Standing Eye Height Range (cm)	Standing Elbow Height Range (cm)
152	41	109	58	145	99
155	41	112	58	147	99
157	41	112	61	150	102
160	41	114	61	152	102
163	41	117	63	155	104
165	43	119	63	157	107
168	43	119	66	160	109
170	44	122	66	162	109
173	45	124	67	161	112
175	45	127	68	165	114
178	45	127	69	168	114
180	45	130	71	170	117
183	46	132	71	174	119
185	46	132	73	175	119
188	48	135	74	178	122
191	51	137	74	180	124
193	51	140	76	183	127

Squarespace and Weebly allow for embedding through code blocks or embed code sections, respectively. Most website editors support adding custom HTML or embed codes, making it easy to integrate the calculator widget. It's important to note that our widgets are provided as-is, and Inch Calculator reserves the right to suspend availability without prior notice.

The HTML CODE:

```
<div id="inch-calculator-icw" data-ct="desk_height"
  data-cw="100%" data-ch="875" data-cv="MTE3MjA3MzU0OTM="><div
  id="inch-calculator-icwh">Desk Height Calculator</div><div
  id="inch-calculator-icwf"><a id="inch-calculator-icwi" href="
  https://www.inchcalculator.com/desk-height-calculator/"
  target="_blank"><img id="inch-calculator-icwl" src="https://
  cdn.inchcalculator.com/e/inch-calculator-logo-tiny.png"
  alt="Inch Calculator Logo" width="40" height="49"><span
  id="inch-calculator-icwb">Inch Calculator</span></a></div>
  </div><script src="https://cdn.inchcalculator.com/e/widgets.
  min.js" async defer></script>
```

The provided HTML code snippet integrates the Desk Height Calculator widget from Inch Calculator into a webpage. The widget is encapsulated within a <div> element with the ID "inch-calculator-icw." It includes attributes such as data-ct (calculator type), data-cw (width), data-ch (height), and data-cv (a hashed identifier). Inside this container, there are child elements: "inch-calculator-icwh" for the widget title ("Desk Height Calculator") and "inch-calculator-icwf" for a footer section containing a link back to Inch Calculator's website. This footer includes an image logo ("inch-calculator-icwl") and a text link ("inch-calculator-icwb"). The script tag at the end loads the necessary JavaScript file from Inch Calculator's CDN asynchronously to render the widget on the webpage.

- **Height percentile calculators** in ergonomic design software are essential for assessing human stature across various populations and age groups. By using statistical data, these tools determine an individual's height percentile within their demographic, aiding in designing products and environments that accommodate diverse users. The calculators compare a person's height to general or specific demographics, providing crucial data for accurate digital human representations. Relying on CDC and WHO data, particularly for children, these tools support both imperial and metric units. This flexibility ensures precise modeling, enhancing authenticity in virtual environments, educational tools, and healthcare applications (https://www.gigacalculator.com/calculators/height-percentile-calculator.php) [9]. Gigacalculator offers a free Height Percentile Calculator that can be easily embedded into websites, providing a valuable tool for designing digital humans with accurate anthropometric data. Website owners can integrate the calculator using straightforward HTML code provided by Gigacalculator. This embeddable feature allows designers and developers to incorporate precise height percentile calculations directly into their digital platforms, enhancing the authenticity and realism of virtual environments, simulations, and educational tools. By leveraging the Height Percentile Calculator, designers can ensure that their digital human representations reflect real-world demographics and anthropometric norms, catering to diverse user bases and promoting inclusivity in digital experiences. This accessibility supports various applications, including personalized health assessments, ergonomic design considerations, and educational simulations, thereby optimizing user engagement and interaction in virtual contexts.

Helping digital human design, integrating a Height Percentile Calculator widget from Gigacalculator onto user website is straightforward. Begin by selecting the Height Percentile Calculator by name from Gigacalculator's site. Customize the widget's dimensions and title to match user website's aesthetics and needs. Once tailored, Gigacalculator provides the HTML code for embedding. Simply copy and paste this code into user's website HTML where the user wants the Height Percentile Calculator to appear. This seamless process empowers users with a valuable tool for calculating height percentiles, enhancing engagement by offering easy access to anthropometric insights directly on the user site. Such integration not only enhances user experience but also supports various digital applications, including healthcare, education, and virtual simulations.

Widget width: 450 / Widget title: Height Percentile Calculator

HTML Code

```
<div class="gigacalculator" data-tool="/calculators/height-
  percentile-calculator.php" data-width="450">
 <div class="gigacalctitle">Height Percentile Calculator</div>
 <div class="gigacalcfooter">
 <a href="https://www.gigacalculator.com/calculators/height-
  percentile-calculator.php">Height Percentile Calculator</a>
  by <a class="gigacalclink" href="https://www.gigacalculator.
  com/">GIGAcalculator.com</a>
 </div>
</div>
<script defer src="https://cdn.gigacalculator.com/embed.min.
  js"></script>
```

The HTML code provided is for embedding the Height Percentile Calculator from Gigacalculator.com into a website. Here's a breakdown of the code:

- **`<div class="gigacalculator" ...>`**: This div contains attributes (`data-tool` and `data-width`) that specify the tool URL (`/calculators/height-percentile-calculator.php`) and the width of the embedded calculator (`450` pixels).
- **`<div class="gigacalctitle">...</div>`**: Displays the title "Height Percentile Calculator" within the embedded widget.
- **`<div class="gigacalcfooter">...</div>`**: Includes a footer with links to the original calculator page and Gigacalculator.com.
- **`<script defer src="`**https://cdn.gigacalculator.com/embed.min.js**`"></script>`**: Loads an external JavaScript file (`embed.min.js`) from Gigacalculator.com's content delivery network (CDN). This script is responsible for embedding the calculator functionality into the webpage.

To embed this calculator on the user's website, simply place this HTML code where the user wants the calculator to appear. Adjust the `data-width` attribute to change the width of the embedded calculator as needed. This setup allows visitors to use the Height Percentile Calculator directly on the user site, enhancing user engagement and functionality.

Height Percentile Calculator Calculation examples (Metric system) [9]:

Example 1 parameters: Height: 185 cm; Gender: Male; Age: 42; From: United States.

Example 2 parameters: Height: 171 cm; Gender: Female; Age: 40; From: United States.

Calculation results:

Example 1 results: Your height is in the top: 8th percentile*; You are taller than: 92.31% *; You are shorter than: 7.69% *; * of adult males from the United States.

Example 2 results: Your height is in the top: 6th percentile*; You are taller than: 93.73% *; You are shorter than: 6.27% *; * of adult females from the United States.

- **BMI calculator** helps in assessing the body composition and weight status of individuals interacting with ergonomic designs. By calculating BMI using height and weight, designers obtain a standardized metric to evaluate health risks associated with different weight categories, including underweight, normal weight, overweight, and obesity. Integrating BMI data into 3D digital models allows for the optimization of product dimensions, seating, and workspace layouts, promoting ergonomic comfort and healthy behaviors. This integration enhances the creation of realistic and diverse avatars that accurately represent various body types, improving authenticity in virtual environments, gaming, virtual reality, and simulations. BMI-based customization also supports personalized health and fitness applications, ensuring digital humans are both visually accurate and functional. This highlights the importance of incorporating accurate health metrics into digital human design, ultimately contributing to more lifelike and user-centric digital representations across various applications [10]. Gigacalculator BMI calculator can be easily embedded into websites, allowing users to calculate their BMI directly on the site. This tool provides personalized health insights by displaying the user's BMI category, BMI Prime index, and how much they deviate from the optimal weight range. Designed for adults, it enhances user engagement by offering valuable health information. Integrating accurate BMI data in digital human design improves the realism and functionality of virtual representations, promoting health awareness. This tool also supports the creation of diverse body types in virtual environments, making it a valuable resource for personalized health and fitness applications. Table 10.2 "BMI categories and corresponding BMI prime and risk levels." For instance, the normal (healthy) BMI range of 18.5 to 24.9 kg/m^2, with a BMI Prime of 0.74 to 1.0, indicates very low risk, emphasizing its importance in maintaining optimal health.

How to Get a BMI Calculator Widget?

It's quite straightforward: First, select the BMI calculator (https://www.gigacalculator.com/calculators/bmi-calculator.php) by searching for it by name on the Gigacalculator website. Next, customize the widget's dimensions and title to fit to website's design and requirements. Finally, copy the provided code and place it in the website's HTML to seamlessly integrate the BMI calculator. This easy process allows users to offer a valuable health tool to their visitors, enhancing user engagement and providing personalized health insights directly on the user site [11].

TABLE 10.2
BMI Categories and Corresponding BMI Prime and Risk Levels [10]

Category	BMI Range (kg/m²)	BMI Prime	Risk Level
Very severely underweight	less than 15	less than 0.60	Moderate
Severely underweight	15–15.9	0.60–0.64	Moderate
Underweight	16–18.4	0.64–0.74	Moderate
Normal (healthy)	18.5–24.9	0.74–1.0	Very low
Overweight	25–29.9	1.0–1.2	Low
Moderately obese	30–34.9	1.2–1.4	Moderate
Severely obese	35–39.9	1.4–1.6	High
Very severely obese	40+	1.6+	Very high

Widget width: 450 / Widget title: BMI calculator

```
HTML Code
<div class="gigacalculator" data-tool="/calculators/bmi-
  calculator.php" data-width="450">
 <div class="gigacalctitle">BMI calculator</div>
 <div class="gigacalcfooter">
 <a href="https://www.gigacalculator.com/calculators/
  bmi-calculator.php">BMI calculator</a> by <a
  class="gigacalclink" href="https://www.gigacalculator.
  com/">GIGAcalculator.com</a>
 </div>
</div>
<script defer src="https://cdn.gigacalculator.com/embed.min.
  js"></script>
```

The provided HTML code snippet is for embedding a BMI calculator widget from Gigacalculator.com into a website. Here's a breakdown of what each part of the code does:

- **`<div class="gigacalculator" …>`**: This div contains attributes (`data-tool` and `data-width`) that specify the tool URL (`/calculators/bmi-calculator.php`) and the width of the embedded calculator (`450` pixels).
- **`<div class="gigacalctitle">BMI calculator</div>`**: Displays the title "BMI Calculator" within the embedded widget.
- **`<div class="gigacalcfooter">…</div>`**: Includes a footer with links to the original calculator page and Gigacalculator.com.
- **`<script defer src="https://cdn.gigacalculator.com/embed.min.js"></script>`**: Loads an external JavaScript file (`embed.min.js`) from Gigacalculator.com's content delivery network (CDN). This script is responsible for embedding the calculator functionality into the webpage.

To embed this BMI calculator on the website, simply place this HTML code by personal wish for the calculator to appear. Adjust the `data-width` attribute to change the width of the embedded calculator as needed. This setup allows visitors to use the BMI calculator directly on the user site, providing a valuable health tool and enhancing user engagement with personalized health insights.

BMI Calculation examples (Figure 10.5; Metric system) [10]:

Example 1 parameters: Height 187 cm; Weight 95 kg.
Example 2 parameters: Height 173 cm; Weight 54 kg.
Example 1 results: BMI: 27.17; BMI category: Overweight; BMI Prime: 1.09; Over/under limit: 8.7% over the upper limit; Hazard ratio for all-cause mortality (men)*: 1.12; Hazard ratio for all-cause mortality (women)*: 1.08.
Example 2 results: BMI: 18.04; BMI category: Underweight; BMI Prime: 0.72; Over/under limit: 2.5% under the lower limit; Hazard ratio for all-cause mortality (men)*: 1.83; Hazard ratio for all-cause mortality (women)*: 1.53.

BMI 3D online calculators are invaluable tools for 3D digital design in ergonomics and human factors, allowing users to calculate their BMI and visualize their body shape in three dimensions. Platforms like **BMI Visualizer** (https://www.bmivisualizer.com) enable users to see how their BMI, calculated from height and weight, correlates with body fat percentage and overall body shape. **BMI 3D** (https://www.bmi3d.com/calculator.html) provides insights into weight categories such as normal weight, overweight, or obese based on BMI. The **Biri platform** (https://biri.org/wellness/bmi/) integrates BMI calculations with basal metabolic rate (BMR), offering a comprehensive look at daily energy expenditure at rest.

A very detailed 3D + math data tool is **Meshcapade Me Body Visualizer** (https://bodyvisualizer.com), which also offers a desktop app and allows users to edit and visualize body proportions in detail, making it a versatile tool for ergonomic and human factors design. These platforms help designers create more accurate and inclusive digital human models by incorporating scientifically-backed anthropometric data [12–16]. Figure 10.1 shows the default interface of Meshcapade Me Bodyvisualizer (online version). The first opening of the 3D BMI calculator offers a Desktop App installation option, whose interface is shown in Table 10.3 (3D female model) and Table 10.4 (male model), where 3D digital design of human models is described in detail.

The Meshcapade Me Bodyvisualizer BMI calculator is a highly scientific tool developed based on research by a group of experts who have published detailed information related to the software. The platform strictly follows scientific methods to process received data, enabling the digital generation of 3D geometry based on real models. The open-access scientific articles and projects are available at the following links:

It should be noted that, apart from the purely scientific nature of the developed software, the online platform and the application have a lot of 3D interactive tools and libraries, which significantly support the visualization of the digitally generated model by parameters.

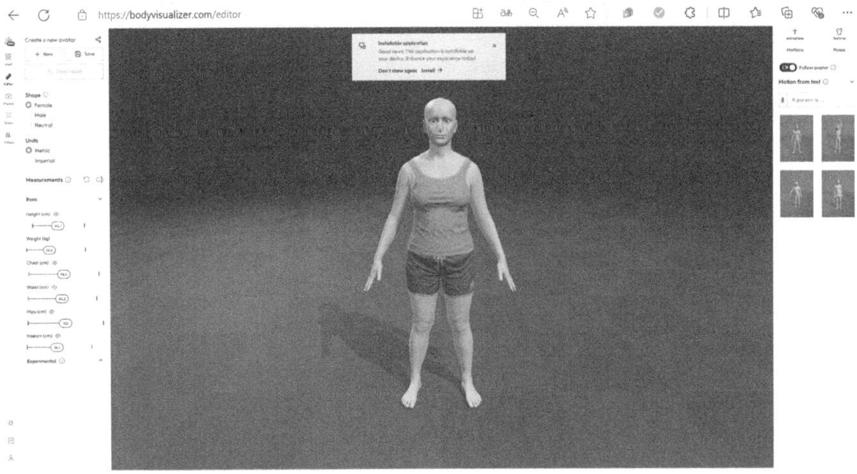

FIGURE 10.1 Meshcapade Me Bodyvisualizer Editor Online BMI calculator [12].

TABLE 10.3
A 3D Digital Design of Human Female Model in Desktop Meshcapade Me Body Visualizer Version

BMI parameters Female Units: Metric Green zone: Normal parameters Red zone: Problem parameters	Meshcapade Me Bodyvisualizer BMI calculator Desktop software 3D Model: f_asian_03_cargopants_a_6_tank_a_4_ALB
Height (cm): 171 cm - red zone <> Weight (kg): 53 – red zone Default parameters of: Chest, Waist, Hips, Inseam	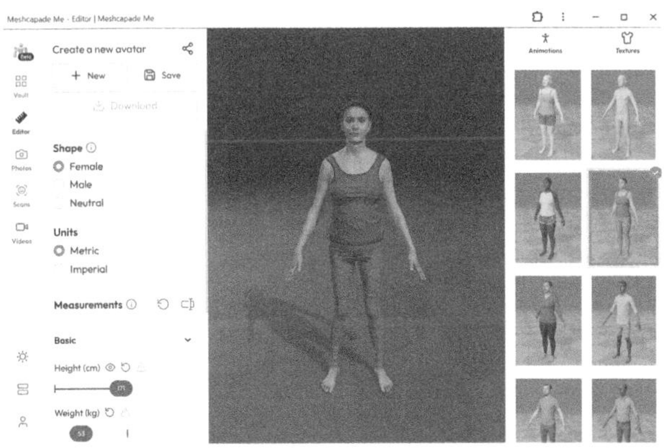

(Continued)

TABLE 10.3
(Continued)

BMI parameters Female Units: Metric Green zone: Normal parameters Red zone: Problem parameters	**Meshcapade Me Bodyvisualizer BMI calculator Desktop software 3D Model: f_asian_03_cargopants_a_6_tank_a_4_ALB**
Height (cm): 171 cm - green zone <> Weight (kg): 76 – green zone Default parameters of: Chest, Waist, Hips, Inseam	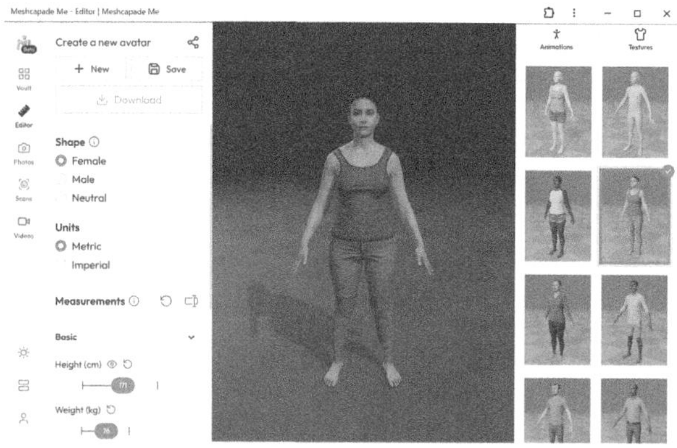
Height (cm): 171 cm - green zone <> Weight (kg): 89 – red zone Default parameters of: Chest, Waist, Hips, Inseam	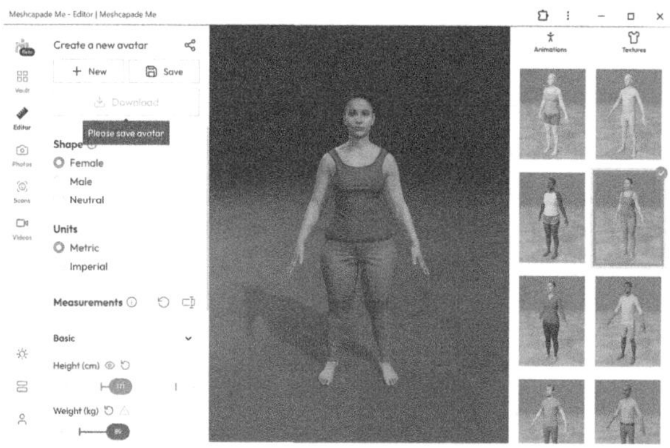
Height (cm): 171 cm - green zone <> Weight (kg): 76 – green zone Increased parameters of: Chest, Waist, Hips, Inseam	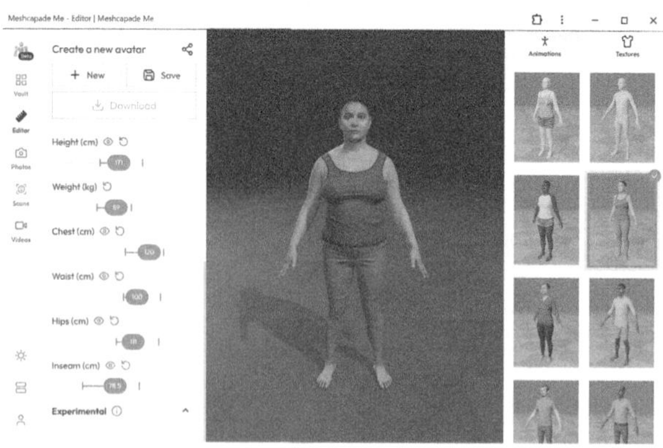

TABLE 10.4
A 3D Digital Design of Human Male Model in Desktop Meshcapade Me Body Visualizer Version

BMI parameters Male Units: Metric Green zone: Normal parameters Red zone: Problem parameters	Meshcapade Me Bodyvisualizer BMI calculator Desktop software 3D Model: m_white_05_slacks_b_12_tshirt_a_15-thumbnail-without-lineart
Height (cm): 187 cm - green zone <> Weight (kg): 46 – red zone Default parameters of: Chest, Waist, Hips, Inseam	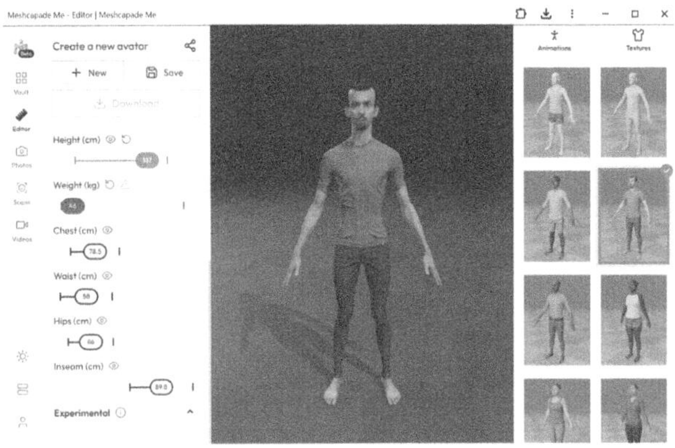
Height (cm): 187 cm - green zone <> Weight (kg): 89 – green zone Default parameters of: Chest, Waist, Hips, Inseam	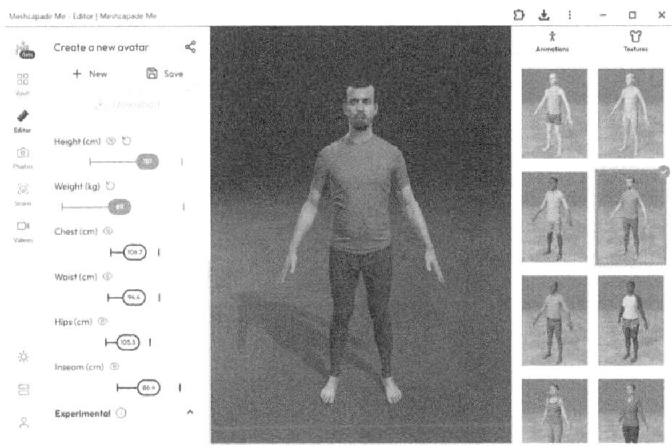
Height (cm): 187 cm - green zone <> Weight (kg): 139 – red zone Parameters of: Chest (136.7 cm), Waist, Hips, Inseam	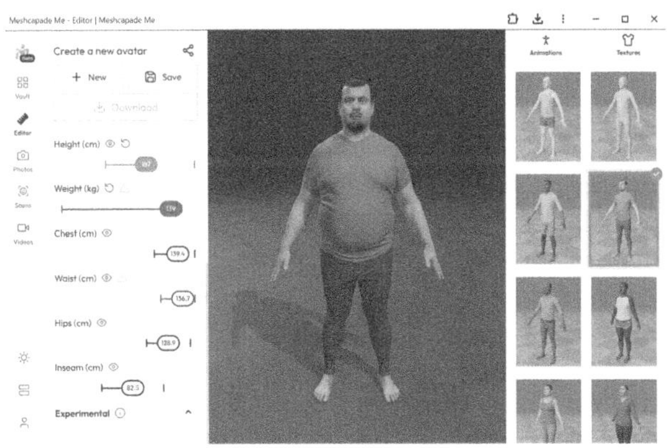

(Continued)

TABLE 10.4
(Continued)

BMI parameters Male Units: Metric Green zone: Normal parameters Red zone: Problem parameters	**Meshcapade Me Bodyvisualizer BMI calculator Desktop software 3D Model: m_white_05_slacks_b_12_ tshirt_a_15-thumbnail-without-lineart**
Height (cm): 187 cm - green zone <> Weight (kg): 89 – green zone Increased parameters of: Chest, Waist, Hips, Inseam	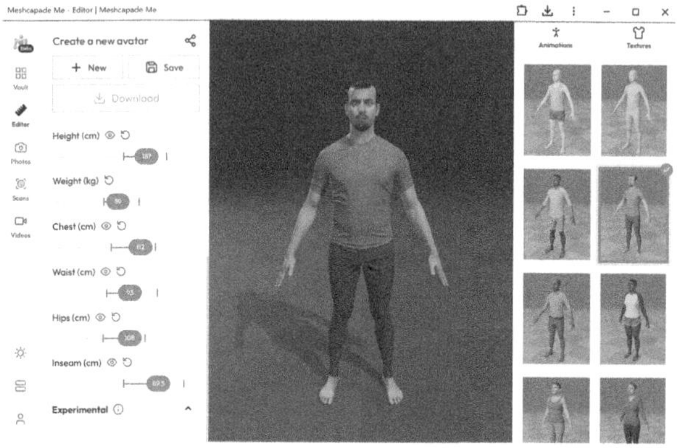

In panel animation, a motion section is available (left), where 3D animated digital models of people are moved in real time. At the same time, in the Textures panel, there is a set of digitally obtained people from real analogs (right) (Figure 10.2). The system allows downloading the graphic files in an expansion.

Moreover, the software allows users to choose poses and positions of their hands (Figure 10.3).

Despite the advantages of the above-mentioned online platforms and the direct connection of the Meshcapade Me Bodyvisualizer BMI calculator application, they are related to the sustainability and support of the system operators and ITs, who deliver quality to the developed applications. It should be noted that if online platforms evolve, change, resell, stop functioning for technical reasons, or are abandoned, as well as other reasons, then it is good to provide stable OS installation software for calculating BMI. In the context of this, some of the free-to-use BMI calculators that are resistant to use are BMI calculators for men, women, and kids [17–19].

A notable software is a **Fitness Calculator Module** (https://www.softpedia.com/get/Others/Home-Education/Fitness-Calculator-Module.shtml) software, which can be used parallel for real body checking and measurement in the development of 3D digital human models. Obtaining data by calculating BMI is useful for athletes and fitness enthusiasts, as well as people adhering to a healthy and active lifestyle. The Fitness Calculator Module is a sophisticated software solution designed to provide a variety of parameters essential for fitness workouts and basic health information. This user-friendly application installs quickly and features a clean, intuitive graphical interface with numerous practical tools. It enables users to calculate their maximum heart rate based on age and resting heart rate by counting beats per minute.

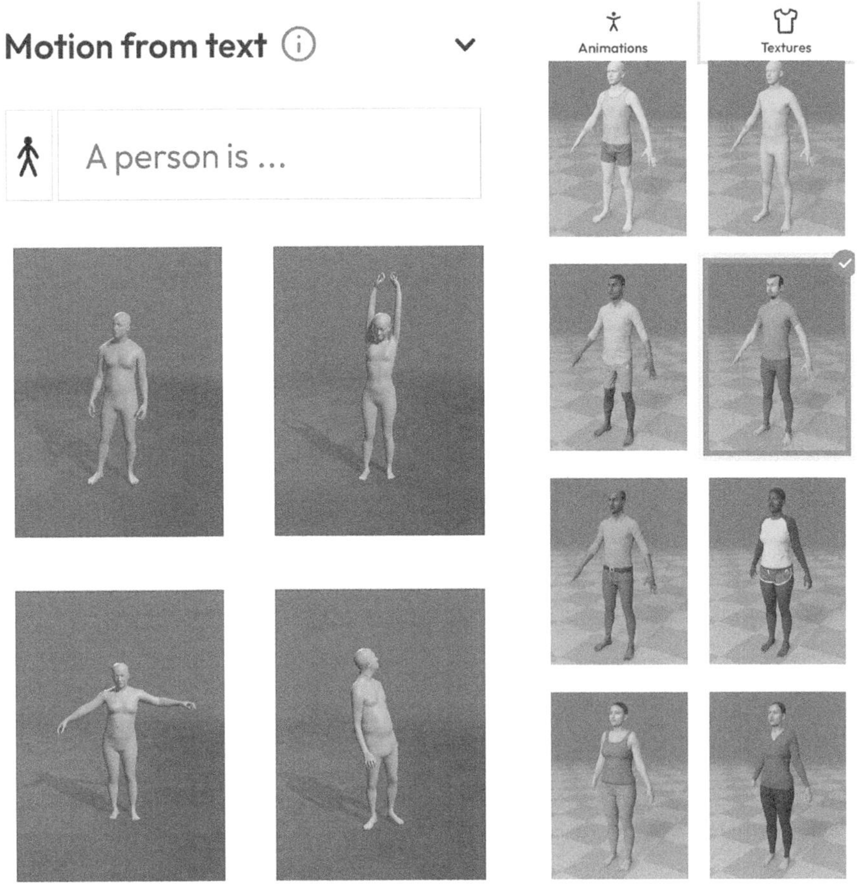

FIGURE 10.2 Meshcapade Me Bodyvisualizer BMI calculator - Motion. Left: Motion from text; Right: Animations and Textures [12].

Additionally, it offers insights into the heart rate zone for aerobic exercises. Users can easily determine their BMI by entering weight and height data, and calculate their BMR, indicating the calories needed to maintain bodily functions at rest. The software also provides calculations for allowable calorie intake, waist-to-hip ratio, ideal weight, fat range, body mass, and daily protein requirements, including metric conversions. Despite its comprehensive features, it lacks the ability to export data to a text file. Table 10.5 shows a fitness study using free software for Windows OS. The table examines the overall human condition. Calculations with Fitness Calculator Module free software covering: Maximum Heart Rate; Resting Heart Rate; Heart Rate Zone; BMI; BMR; Total Daily Energy Expenditure; Waist Hip Ratio; Ideal Weight Range; Ideal Fat Range; Lean Body Mass; Daily Protein Requirement; and Metric Conversions. This software is particularly valuable in the realm of 3D digital design in ergonomics and human factors, where accurate health data is crucial for digital health analysis, developing health programs, and monitoring health metrics.

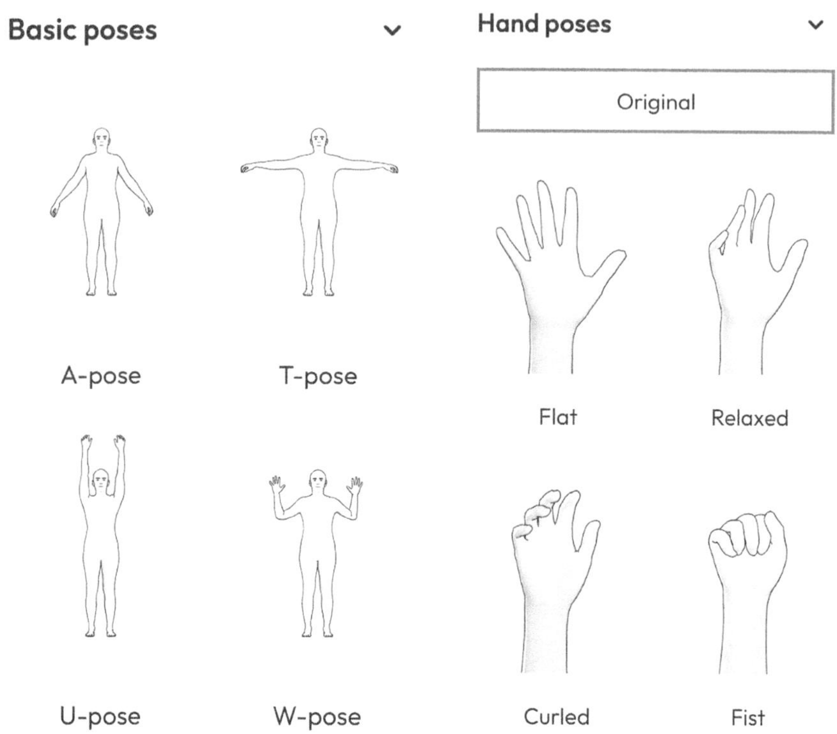

FIGURE 10.3 Meshcapade Me Bodyvisualizer BMI calculator. Left: Basic poses; Right: Hand poses [12].

By integrating these parameters, designers can enhance interactive 3D digital ergonomics and bioengineering applications, ensuring they are tailored to real human physiological data. This integration aids in creating more effective and personalized ergonomic designs, improving both user comfort and safety. The Fitness Calculator Module thus serves as an essential tool for designers, enabling them to incorporate detailed health metrics into their software, fostering a more detailed approach to digital ergonomics and human factors engineering [20].

BMR Health Calculator is a user-friendly software application designed to calculate BMI and BMR [21]. Its approachable options cater to users with no prior experience in health calculators. The setup process is quick and straightforward, free from unfamiliar options, prerequisite software, or third-party components. The interface features a standard window with a well-organized layout, displaying a BMI graph. Users can input information in either the English or metric system in separate panes for age, weight, and height for both males and females, with calculations performed instantly as long as values are within specified ranges. The results are displayed via balloon messages. Figure 10.4 shows an example of the calculation of BMI and BMR of males.

TABLE 10.5
Fitness Calculations Study Using Free Software Fitness Calculator Module for Windows OS [20].

Calculations:	Fitness Calculator Module: Software interface and display calculation results
Calculation of Maximum Heart Rate; Input data of age (y): 26; Result: Maximum Heart Rate is 194	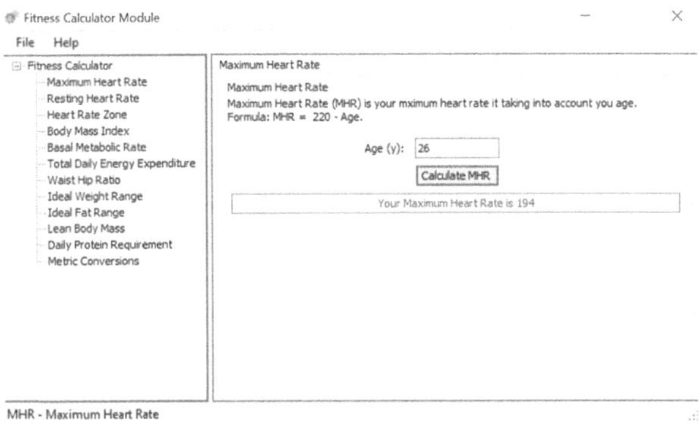
Calculation of Resting Heart Rate; Input data of Pulse (10 seconds): 8; Result: Resting Heart Rate is 48	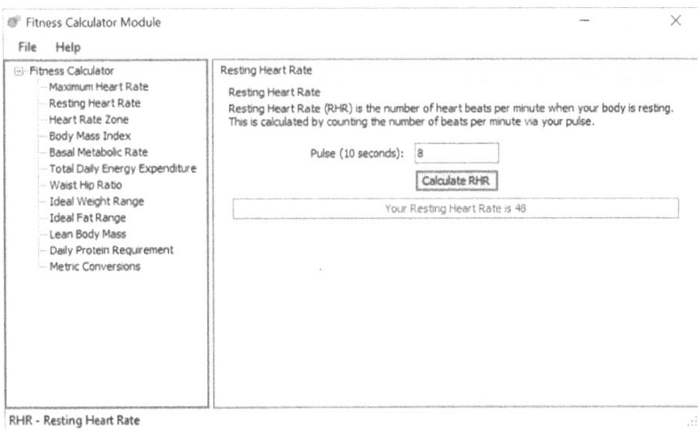
Calculation of Heart Rate Zone; Input data of MHR: 185 and RHR: 52; Result Formula 1: Low Fitness: Between 118,5 and 131,8; Average Fitness: Between 131,8 and 145,1; High Fitness: Between 151,75 and 165,05. Formula 2: Range: Between 120,25 and 157,25	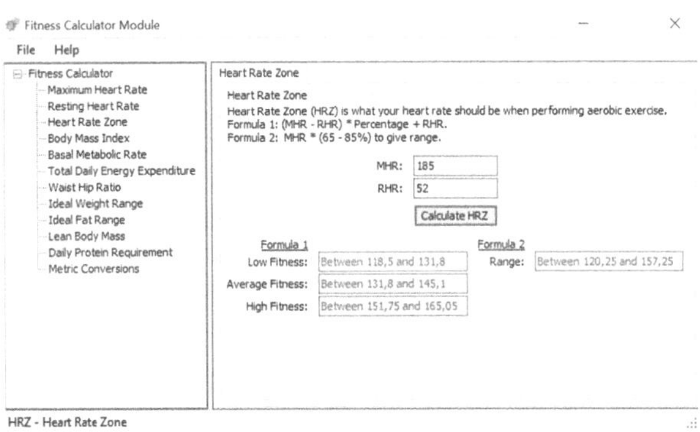

(Continued)

TABLE 10.5 (Continued)

Calculations:	Fitness Calculator Module: Software interface and display calculation results
Calculation of Body Mass Index; Input data of BMI: Weight (kgs): 89; Height (m): 1,87. Result: BMI index is 25,45.	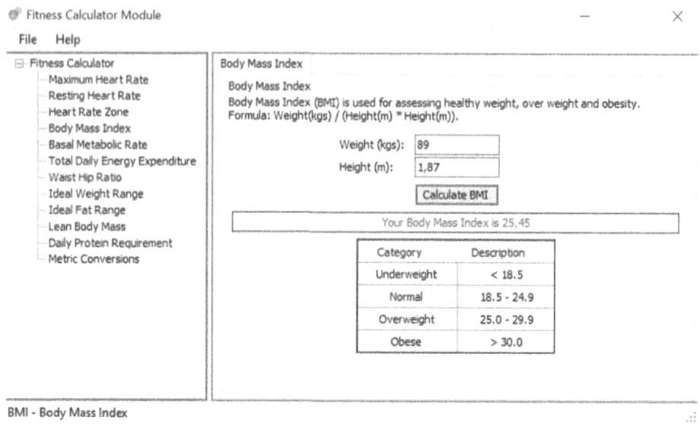
Calculation of Basal Metabolic Rate; Input data of BMR: Weight (kgs): 89; Height (cm): 187; Age (y): 26; Gender (M/F): Male. Result: Basal Metabolic Rate is 2043,5	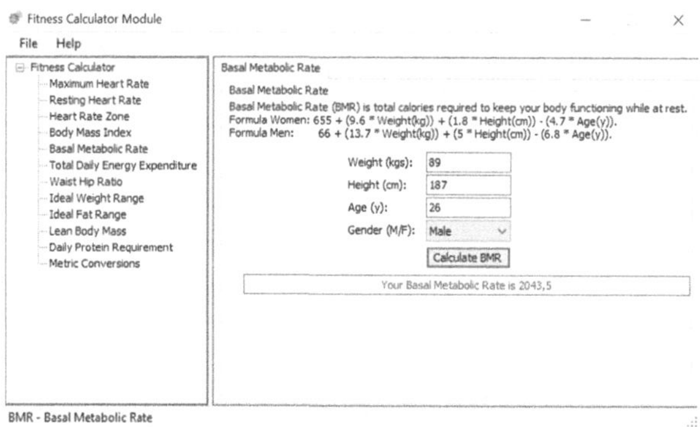
Calculation of Total Daily Energy Expenditure; Input data of BMR: 2043,5	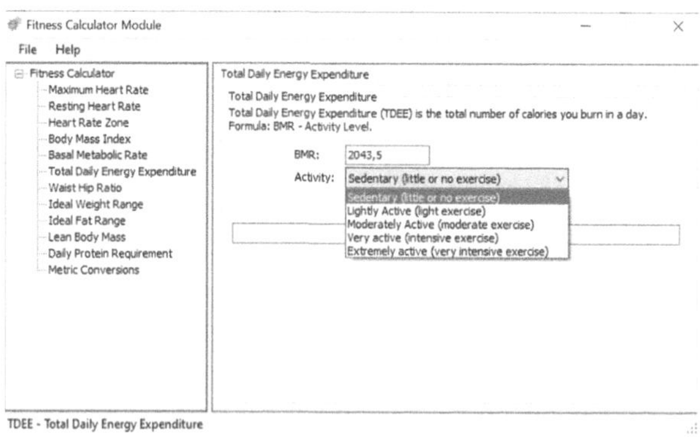

Activity:	**TDEE Result**:
Sedentary (little or no exercise)	2452,2
Lightly Active (light exercise)	2809,81

(Continued)

TABLE 10.5
(Continued)

Calculations:	Fitness Calculator Module: Software interface and display calculation results	
	Moderately Active (moderate exercise)	3167,42
	Very active (intensive exercise)	3525,04
	Extremely active (very intensive exercise)	3882,65

Calculation of Waist Hip Ratio; Input data of WHR: Waist (inches): 48; Hips (inches): 43. Result: Waist to Hip ratio is 1,12. Reference Waist and Hip data: https://www.catrade.ch/wsshop/Catrade/img/4_SizeGuide/SizeGuide_Prolimit.pdf

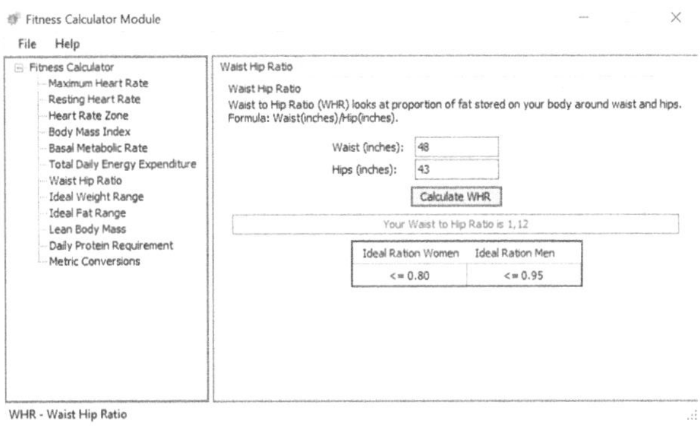

Calculation of Ideal Weight Range; Input data of IWR: Height (m): 1,87. Result: IWR is between 64,69 and87,07

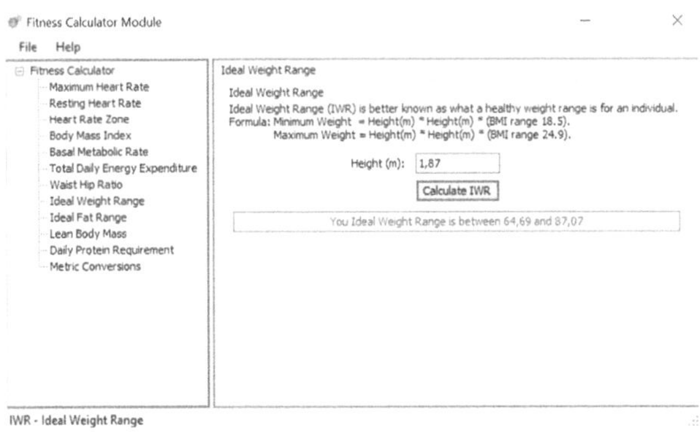

Ready Guide to Ideal Fat Range in %;

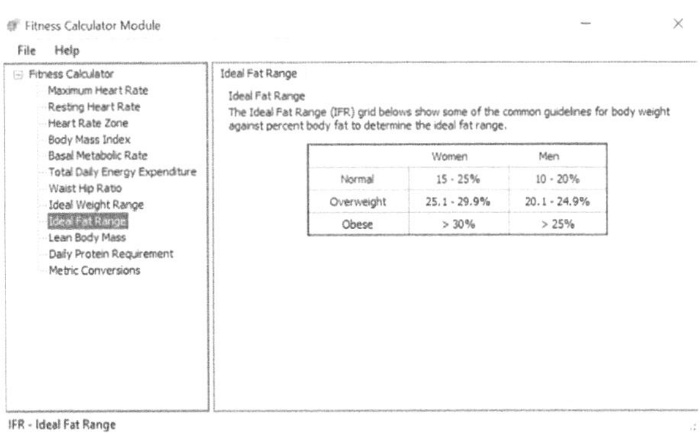

(*Continued*)

TABLE 10.5
(Continued)

Calculations:

Fitness Calculator Module: Software interface and display calculation results

Calculation of Lean Body Mass; Input data of LBM: Weight (kgs): 89; Fat (%): 25. Result: LBM is 66,75 (kgs) and body fat is 22,25 (kgs) Reference: https://www.spidersport.com/lean-body-mass_en.php

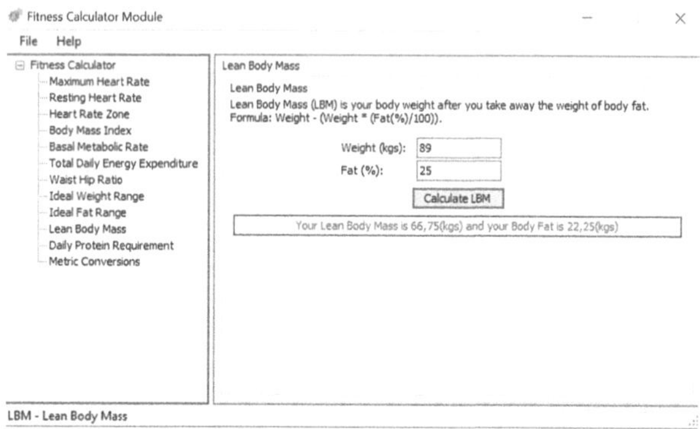

Calculation of Daily Protein Requirement; Input data of DPR: LBM (pounds): 147,18

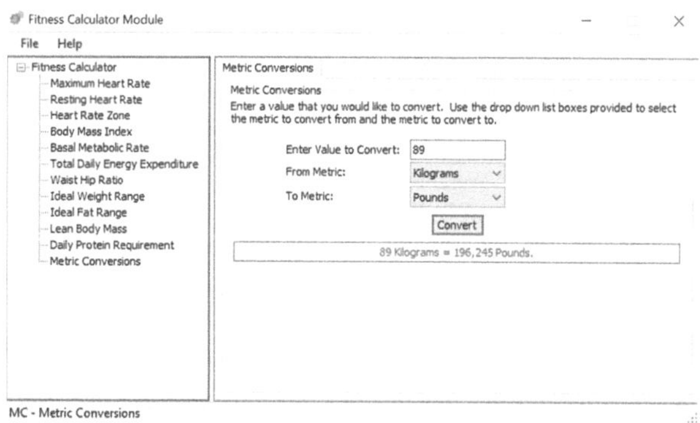

Training Factor:	**TPR Result (grams)**:
No training	73,59
Light training	88,31
Moderately training 3 times per week	103,03
Moderate daily training	117,74
Heavy training	132,46
Heavy training daily	147,18
Heavy training daily and cardio 3 times per week	220,77

(Continued)

TABLE 10.5
(Continued)

Calculations:	**Fitness Calculator Module: Software interface and display calculation results**
Calculation of Metric Conversions. Input data of MC: Value to Convert: 89 from Metric kilograms to Metric Pounds. Result: 89 Kilograms = 196,245 Pounds	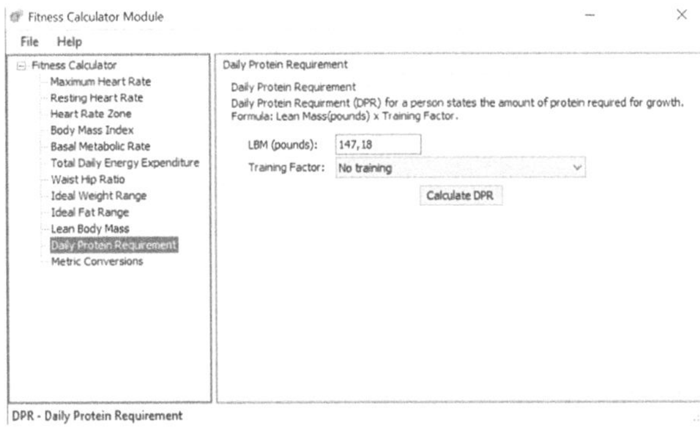

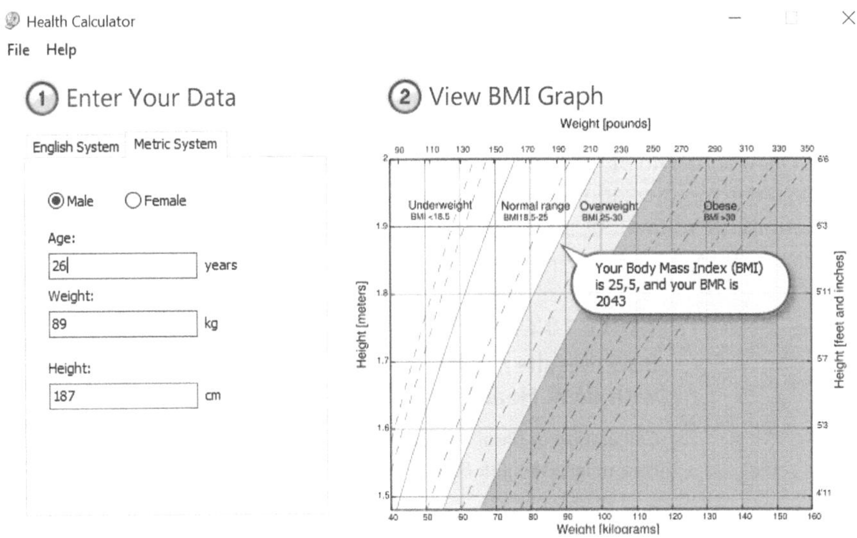

FIGURE 10.4 Calculation of BMI and BMR; Input data of BMI: male, Metric; 26-year-old, Weight: 89; Height: 187. BMR Result is: 2043 [21].

Table 10.6 shows the BMR results of men with BMI data Metric; 26, 31, 43, 55, 62, 71, and 80 years old, Weight: 89; Height: 187.

The **BMI Chart Calculator software** is well graphically illustrated, making it easy to use for calculating the BMI based on height and weight [22]. This straightforward program provides a graphical representation of the BMI chart, allowing users to see how much weight they should lose or gain to fall within the normal, healthy range. Installing the tool is quick, provided the user has the .NET framework installed to run the weight loss calculator. The interface features a large, simple window displaying a BMI table with weights, heights, and corresponding BMIs, color-coded for easy interpretation. Users can input their height and weight using either the imperial system ("Standard" tab, with feet, inches, and pounds) or the metric system ("Metric" tab, with centimeters and kilograms). After clicking the "Calculate" button, the app instantly determines the user's BMI and range, pinpointing it on the chart with a black box to show the user's current status. The "Reset" button clears values for new calculations. Figure 10.5 shows an example of the calculation of BMI with values and a graphical representation.

Related to 3D digital design in ergonomics and human factors, the presented software and tools in this section are invaluable. They support the development of ergonomic designs and health programs by providing accurate, real-time data. This integration enables designers to create more interactive and human-centric digital applications and software, improving user experience and safety. The use of open-source and free initiatives and collaborative repositories promotes interdisciplinary collaboration among ergonomists, engineers, healthcare professionals, and software developers. This fosters innovation and continuous improvement in ergonomic design methodologies, ensuring that digital human modeling is both accurate and reflective of real-world human factors. The incorporation of such comprehensive health calculators into 3D digital design platforms not only enhances the accuracy of digital human representations but also bridges the gap between digital and real-world applications. It allows for more precise health monitoring, the development of

TABLE 10.6
BMR Results of Men (Metric) Calculated with BMR Health Calculator

Calculation of Basal Metabolic Rate (BMR). Parameters: Weight: 89 Height: 187 BMI: 25,5

Age	BMR
26	2043
31	2009
43	1928
55	1846
62	1799
71	1737
80	1676

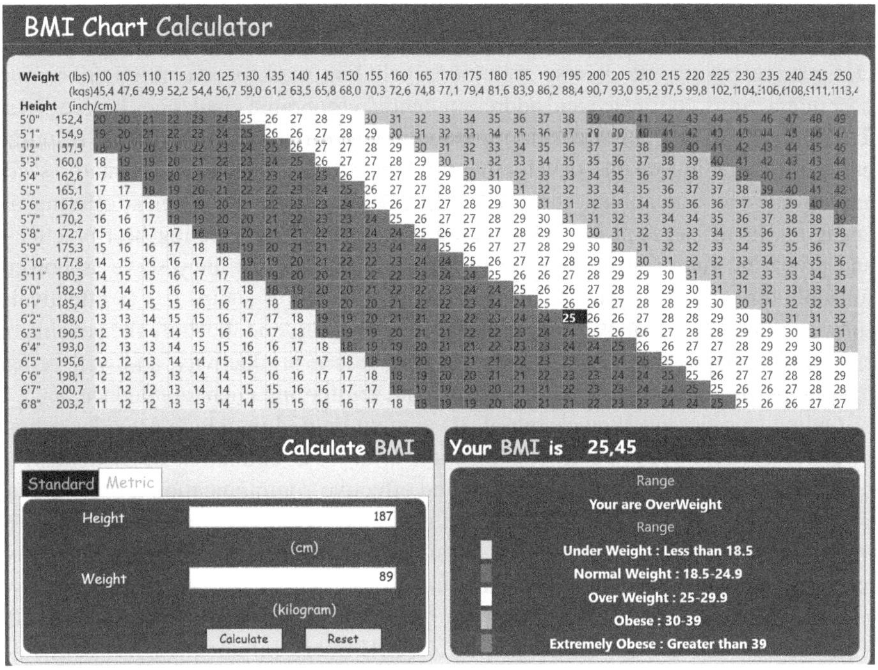

FIGURE 10.5 Calculation of BMI. Input data of BMI: Metric; Height: 187; Weight: 89; BMI Chart Calculator software shows BMI result is: 25,45.

personalized health programs, and the improvement of ergonomic and bioengineering designs. This detailed approach ultimately contributes to better health outcomes and more efficient, human-centered design solutions.

10.3 3D DIGITAL INTERACTIVE PRESENTATION

There are lots of platforms and applications for 3D digital interactive presentations, which support visualization technologies to craft virtual environments where stakeholders, including clients, end-users, and multidisciplinary teams, can explore and interact with digital models of products, workspaces, and ergonomic interventions in real time. One of the standout benefits of this approach is its capacity to translate complex ergonomic principles and design concepts into highly intuitive, engaging formats. By utilizing real-time rendering, designers can create realistic simulations that vividly illustrate ergonomic features like adjustable furniture, anthropometric considerations, and user interfaces, allowing stakeholders to navigate through virtual environments, manipulate objects, and experience ergonomic solutions as if they were physically present. This immersive experience not only clarifies the intricacies of ergonomic design but also enhances collaborative efforts by enabling real-time feedback and iterative design processes. Features such as interactive annotations, animations, and ergonomic assessments streamline communication, ensuring a shared understanding of design objectives and requirements while fostering innovation and refinement. From a practical perspective, 3D digital interactive presentations significantly boost the feasibility and cost-effectiveness of ergonomic design evaluations. Designers can simulate diverse scenarios, including workspace configurations and user interactions, to foresee and address potential ergonomic challenges before physical prototyping or implementation, reducing design risks and optimizing resource use while speeding up the development process. The interactive presentations are appropriate for educational and training purposes within ergonomics, offering an engaging, hands-on approach to teaching principles, demonstrating best practices, and preparing future professionals. In industry applications, they serve as powerful tools for marketing and client presentations, enabling companies to effectively showcase ergonomic innovations, highlight key features, and illustrate the benefits of ergonomic designs in enhancing user comfort, safety, productivity, and well-being, thus supporting informed decision-making and reinforcing client confidence.

Online sharing and showcasing of 3D content on p3d.in, which provides robust tools for communication and visualization in any 3D animation or visualization project, is crucial for maintaining clear and effective communication among digital designers [23]. This service goes beyond mere screenshots and clay renders, offering interactive viewers that make it possible to present the 3D models to clients, teachers, students, or potential customers in a more engaging and informative manner. With support for a wide range of common 3D file formats such as GL Transmission Format (.gltf/.glb), Wavefront OBJ (.obj), Filmbox (FBX) (.fbx), and STL (.stl), p3d.in ensures compatibility with all leading 3D software packages, allowing for easy model uploads. The platform automatically handles intricate details like vertex normals for smooth shading, UV mapping for textures, and aligns models for optimal viewing. It also provides tools for material editing, where users can tweak properties like diffuse color, specular reflection, opacity, and more, to perfect their model's appearance. Users can upload multiple files with sizes up to 100 MB on a free tier and 300 MB on a premium subscription, with special considerations like automating the handling of .mtl files alongside .obj for material and texture information. p3d.in also supports advanced material properties such as metalness, roughness, sheen, clearcoat, and emissive maps, enabling the creation of realistic and detailed models. The platform offers features like Augmented Reality (AR) viewing, allowing users to see models in a real-world context using compatible Android or iOS devices, with models needing to adhere to certain specifications like vertex and face limits, and specific material and texture constraints to be AR-ready. The platform's user-friendly interface and comprehensive support for model editing and AR visualization make it an invaluable tool for anyone looking to present their 3D content in an accessible, professional, and interactive manner [24; Section XI, 15]. The interactive presentation of a 3D digital human male model with parameters Faces: 33.266 and Vertices: 18,462 is available online: https://p3d.in/P6B87/wire. Example source: https://www.ieti.net/TERP/2021V5N2/2.pdf

10.4 ARTIFICIAL INTELLIGENCE AND OTHER APPLICATIONS

Focused on 3D digital design for ergonomics and human factors, the integration of artificial intelligence (AI) to create realistic images and video content is a big advancement. Recently AI has fundamentally changed digital design by enabling the generation of digital images and videos that were previously unattainable. This capability is crucial in ergonomics, where the visual accuracy of models and simulations can significantly impact the effectiveness of ergonomic assessments and interventions. AI algorithms, particularly those based on deep learning and neural networks, can produce detailed and lifelike representations of human figures, environments, and interactions, which are essential for precise ergonomic analysis. By employing AI algorithms to process and enhance images and video data, designers can create dynamic simulations that showcase how ergonomic designs perform over time. This temporal aspect is important for understanding how ergonomic solutions adapt to changing conditions and how users interact with them over extended periods. AI-driven tools can automate the process of creating realistic images and videos, significantly reducing the time and effort required. This efficiency allows ergonomic designers to rapidly iterate on their designs, testing multiple scenarios and refining their solutions based on real-time feedback. The ability to quickly generate and visualize realistic simulations accelerates the design process and enhances the overall quality of ergonomic interventions. AI supports the facilitating collaborative design processes. Realistic images and videos generated by AI can be shared and reviewed by multidisciplinary teams, including engineers, designers, healthcare professionals, and end-users. This collaborative approach ensures that ergonomic solutions are thoroughly vetted and optimized for real-world applications. The use of AI in creating realistic images, videos, and other digital content supports the education and training of ergonomic practitioners. By providing simulations, AI can enhance the learning experience, allowing practitioners to gain a deeper understanding of ergonomic principles and their practical applications.

The book emphasizes the enhanced application of Open Source Blender software, a versatile and actively developed platform that integrates AI tools to revolutionize the design process. Blender's open-source framework allows for the seamless incorporation of AI-driven features, such as AI-powered rendering, generative design, and automated modeling, which significantly improve the efficiency and accuracy of ergonomic simulations. By leveraging Blender's capabilities, designers can create highly realistic 3D models, dynamic simulations, and lifelike visualizations of human interactions with products or environments. The platform's adaptability and continuous development make it an ideal tool for advancing ergonomic design, enabling rapid iteration, collaborative workflows, and innovative solutions that align with human-centered design principles. Among the key tools highlighted are AI Render and 3D AI Studio—Blender Bridge, which significantly enhance the efficiency, creativity, and precision of digital design workflows.

- **AI Render** (https://blender-addons.org/ai-render/) is a powerful add-on that leverages AI to generate images based on text prompts and scenes. It eliminates the need for local code execution, allowing users to create AI-generated visuals without running complex software on their computers. For advanced users, it supports local installation with Stable Diffusion, enabling greater control over the rendering process. One of its standout features is the ability to animate Stable Diffusion settings and prompts, making it possible to create dynamic, AI-driven animations. Additionally, its batch processing capability allows designers to efficiently test various settings and prompts, streamlining the iterative design process. Released under the MIT License (https://github.com/benrugg/AI-Render?tab=MIT-1-ov-file#readme), AI Render is free to use, modify, and distribute, making it accessible to a wide range of users. The tool can be downloaded from its GitHub repository (https://github.com/benrugg/AI-Render).

- **3D AI Studio—Blender Bridge** is another innovative tool (https://www.3daistudio.com/BlenderAddon), which seamlessly connects Blender with 3D AI Studio, a platform for generating AI-created 3D models. This add-on allows users to import AI-generated models directly into Blender with just one click, complete with automatically configured materials and textures. Compatible with Blender 4.0 and newer versions, the Blender Bridge simplifies the process of incorporating AI-generated assets into projects, saving time and effort while maintaining high-quality results. Its seamless integration with 3D AI Studio accounts ensures a smooth workflow, making it an invaluable tool for designers looking to incorporate AI into their creative processes.

Other real-time 3D resources like **BlenderKit** and the **IKEA Browser addon** significantly enhance the design workflow within Blender by providing ready-to-use 3D models and assets. BlenderKit (https://www.blenderkit.com/) is an extensive library seamlessly integrated into Blender, offering over 75,000 assets, including 3D models, materials, brushes, HDRIs, and node groups, with 51% available for free. All assets are licensed for both commercial and noncommercial use, making it a valuable resource for designers. Similarly, the IKEA Browser addon (https://extensions.blender.org/add-ons/ikea-browser/) allows users to browse IKEA's catalog and import 3D furniture models directly into Blender with a single click. These tools streamline the design process by providing high-quality, standardized assets, enabling designers to focus on creativity and ergonomic solutions. However, users must pay attention to licensing terms and prioritize their individual needs when utilizing these resources. Together, these add-ons expand Blender's capabilities, making it a more versatile and efficient platform for 3D design.

These tools not only enhance the visual quality and realism of ergonomic simulations but also streamline the design process, enabling designers to create more effective and human-centered solutions. As AI technology and real-time 3D networks continue to evolve, their applications in Blender and ergonomic design will undoubtedly expand, driving further innovation in the field.

The book "3D Digital Design in Ergonomics and Human Factors" explores the advanced intersection of technology and human-centered design, highlighting one very innovative tool: Craiyon, formerly known as DALL-E mini, an AI-powered online platform accessible at https://www.craiyon.com/. This platform marks a significant advancement in the ability to create realistic images from text prompts, a functionality that is transformative for digital designers needing to visualize complex scenarios, environments, and human interactions without the traditional extensive investment of time and resources. By allowing users to input descriptive text to generate images that closely mimic real-world scenes, Craiyon's ability to facilitate rapid visualization and iteration significantly accelerates the design process, enhancing the precision of digital quality.

REFERENCES

[1] Human Generator Add-On for Blender 3D https://brandon3d.com/human-generator/
[2] Human Generator: Guide https://help.humgen3d.com/guide
[3] Human Generator: Base content https://www.humgen3d.com/base-content
[4] Procedural Crowds add-on for Blender https://brandon3d.com/procedural-crowds-add-on-for-blender-review/
[5] OHCOW: Occupational Health APPS, TOOLS and CALCULATORS https://www.ohcow.on.ca/resources/apps-tools-calculators4/
[6] ErgoFellow 3.0 software https://www.fbfsistemas.com/ergonomics.html
[7] Upliftdesk Ergonomic Calculator https://www.upliftdesk.com/ergonomic-calculator/
[8] Desk Height Calculator Widget https://www.inchcalculator.com/widgets/w/desk-height/
[9] GigaCalculator: Height Percentile Calculator https://www.gigacalculator.com/calculators/height-percentile-calculator.php
[10] GigaCalculator: Body Mass Index Calculator https://www.gigacalculator.com/calculators/bmi-calculator.php
[11] GigaCalculator: Body Mass Index Calculator. How to get a widget? https://www.gigacalculator.com/calculator-widget.php?calcurl=/calculators/bmi-calculator.php#howto
[12] Meshcapade Me Body Visualizer https://bodyvisualizer.com
[13] Meshcapade. https://meshcapade.com/science
[14] Yao Feng, Jing Lin, Sai Kumar Dwivedi, Yu Sun, Priyanka Patel, Michael J. Black. ChatPose: Chatting about 3D Human Pose, CVPR 2024. Project page: https://yfeng95.github.io/ChatPose/; Paper: https://arxiv.org/pdf/2311.18836
[15] Sai Kumar Dwivedi, Yu Sun, Priyanka Patel, Yao Feng, Michael J. Black. TokenHMR: Advancing Human Mesh Recovery with a Tokenized Pose Representation, CVPR 2024. Project page: https://tokenhmr.is.tue.mpg.de/; Paper: https://tokenhmr.is.tue.mpg.de/media/upload/TokenHMR.pdf
[16] Naureen Mahmood, Nima Ghorbani , Nikolaus F. Troje, Gerard Pons-Moll, Michael J. Black. AMASS: Archive of Motion Capture As Surface Shapes, ICCV 2019. Project page: https://amass.is.tue.mpg.de/; Paper: https://files.is.tue.mpg.de/black/papers/amass.pdf
[17] BMI Calculator for MEN https://www.softpedia.com/get/Others/Home-Education/BMI-Calculator-for-MEN.shtml
[18] BMI Calculator for WOMEN https://www.softpedia.com/get/Others/Home-Education/BMI-Calculator-for-WOMEN.shtml
[19] BMI Calculator for KIDS https://www.softpedia.com/get/Others/Home-Education/BMI-Calculator-for-Kids.shtm
[20] Fitness Calculator Module https://www.softpedia.com/get/Others/Home-Education/Fitness-Calculator-Module.shtml
[21] BMR Health Calculator https://www.softpedia.com/get/Others/Home-Education/BMR-Health-Calculator.shtml
[22] BMI Chart Calculator https://www.softpedia.com/get/Others/Home-Education/BMI-Chart-Calculator.shtml
[23] P3D. p3d.in
[24] P3D. Tutorials https://p3d.in/tutorials

Summary

The book "3D Digital Design in Ergonomics and Human Factors" explores the transformative impact of advanced digital technologies on the field of ergonomics, emphasizing the integration of 3D design, digital modeling, and interactive tools to enhance human well-being and system performance. Through a comprehensive examination of various digital technologies and methodologies, the book elucidates how 3D digital design reshapes the way ergonomic solutions are conceived, developed, and implemented. The book presents 3D digital design as a fundamental stage for creating ergonomic solutions. The text delves into the use of additive technologies to capture and replicate physical environments, ensuring that ergonomic assessments and designs are based on accurate and detailed representations of real-world settings. The book highlights the importance of 3D polygonal-mesh modeling and solid modeling [computer-aided design (CAD)] in the design process. These technologies enable the creation of detailed and shaped digital models, allowing for iterative design, testing, and optimization of ergonomic solutions. By providing a deep dive into the functionalities and applications of these modeling techniques, the book illustrates how they contribute to the creation of ergonomic products that meet the diverse needs of users. A significant part of the book is dedicated to the use of Open-Source Software (OSS) in 3D digital design. The text examines the benefits of OSS licenses and their role in fostering innovation and collaboration in the field of ergonomics. Detailed discussions on popular OSS provide practical experience into their application in human-centered design, emphasizing the accessibility and versatility of human generator tools for ergonomic research and practice.

In the context of human-centered design, the book delves into human 3D digital design using tools like Blender and MB Lab, illustrating how these platforms facilitate the creation of realistic and customizable human models for ergonomic analysis. The application of MakeHuman is also discussed, highlighting its contribution to generating anatomically accurate human figures for ergonomic assessments. The book is rich with practical examples that demonstrate the application of 3D digital design in various ergonomic scenarios. From workplace layout optimization to the design of ergonomic tools and devices, these examples provide real-world context and illustrate the tangible benefits of integrating 3D digital technologies in ergonomic practice. The integration of Artificial Intelligence (AI) and real-time networks into 3D digital design is a key focus of the book, showcasing how these technologies are revolutionizing ergonomic solutions. AI-driven tools enable the creation of realistic images, animations, and 3D models in real time, significantly enhancing the efficiency and accuracy of ergonomic simulations. A notable focus is placed on Ergonomics and Motion Capture (MoCap), emphasizing how MoCap technologies are used to capture human movement and behavior, informing the design of ergonomic interventions that enhance safety and efficiency. The intersection of ergonomics and bioengineering is also explored, particularly in the design of inclusive healthcare solutions such as implants, prostheses, and masks, showcasing the critical role of 3D digital design in addressing diverse healthcare needs. The book discusses the importance of additional digital resources and useful extensions (Add-ons), providing insights into how these tools can extend the capabilities of primary software, enabling more sophisticated and customized ergonomic solutions. The text covers digital human calculators for metrics like ergonomics, height percentile, and BMI, underscoring their utility in ergonomic assessments and design. The book emphasizes the significance of 3D digital interactive presentations in communicating ergonomic designs and principles. By leveraging interactive and immersive presentations, ergonomic experts can effectively convey complex concepts, engage stakeholders, and facilitate collaborative decision-making processes.

Index

Pages in *italics* refer to figures, pages in **bold** refer to tables.